The Explore-Before-Explain Guidebook for Science Education

This guidebook uses an *Explore-Before-Explain* instructional sequence to help you facilitate the design of active meaning-making lessons in science.

Author Pat Brown models and breaks down how an *Explore-Before-Explain* approach ensures students' conceptual understandings are constructed primarily on evidence-based experiences in the classroom. From prompting students to ponder patterns, helping them identify cause-and-effect relationships, to focusing on development of their thinking rather than validating ideas, you can use *Explore-Before-Explain* instruction to help your students feel confident in their thinking and become more self-directed learners. Chapters focus on developing your students' conceptual science understanding through the investigation of natural occurrences. Content and module examples are provided, as well as attention to contemporary standards and safety in science. Rather than acting as a prescriptive program, however, this book adds another element to your curriculum design, showing how lessons can and should include critical elements of active sensemaking when designing an *Explore-Before-Explain* sequence. In addition, the author shows the benefits of bringing *Explore-Before-Explain* outside the classroom to create high-quality professional and curriculum-based learning.

This resource is ideal for K-12 science teachers, as well as building administrators who are looking for a practice-oriented and research-based approach for their science curriculum. As a classroom educator, you can use these strategies for leveraging research into hands-on, minds-on activities to promote more robust and equitable learning environments. For leaders, this resource can be used to develop professional learning experiences for sustained departmental improvement.

Patrick Brown is the executive director of STEAM and career education for the Fort Zumwalt School District in St. Charles, Missouri.

Also Available from Routledge Eye on Education
(www.routledge.com/eyeoneducation)

STEM by Design:
Tools and Strategies to Help Students in Grades 4–8 Solve Real-World Problems, 2nd Edition
Anne Jolly

Teaching Climate Change for Grades 6–12:
Activating Science Teachers to Take on the Climate Crisis Through NGSS, 2nd Edition
Kelley T. Lê

Hydropower Efficiency, Grade 4:
STEM Road Map for Elementary School
Carla C. Johnson, Janet B. Walton and Erin E. Peters-Burton

Habitats Local and Far Away, Grade 1:
STEM Road Map for Elementary School
Carla C. Johnson, Janet B. Walton and Erin E. Peters-Burton

Habitats in the United States, Grade K:
STEM Road Map for Elementary School
Carla C. Johnson, Janet B. Walton and Erin E. Peters-Burton

Culturally Responsive and Sustaining Science Teaching:
Teacher Research and Investigation from Today's Classrooms
Elaine V. Howes and Jamie Wallace

The Explore-Before-Explain Guidebook for Science Education

Creating High Quality Lessons for the Classroom and Professional Learning

Patrick Brown

Routledge
Taylor & Francis Group
NEW YORK AND LONDON

Designed cover image: © Shutterstock and NSTA

First published 2025
by Routledge
605 Third Avenue, New York, NY 10158

and by Routledge
4 Park Square, Milton Park, Abingdon, Oxon, OX14 4RN

Routledge is an imprint of the Taylor & Francis Group, an informa business

A co-publication with NSTA Press

© 2025 National Science Teaching Association

The right of NSTA to be identified as the author of this part of the Work has been asserted in accordance with sections 77 and 78 of the Copyright, Designs and Patents Act 1988.

All rights reserved. No part of this book may be reprinted or reproduced or utilised in any form or by any electronic, mechanical, or other means, now known or hereafter invented, including photocopying and recording, or in any information storage or retrieval system, without permission in writing from the publishers.

Trademark notice: Product or corporate names may be trademarks or registered trademarks, and are used only for identification and explanation without intent to infringe.

ISBN: 978-1-032-95716-6 (hbk)
ISBN: 978-1-032-95715-9 (pbk)
ISBN: 978-1-003-58616-6 (ebk)

DOI: 10.4324/9781003586166

Typeset in Palatino
by Apex CoVantage, LLC

Contents

Foreword		*vii*
Acknowledgments		*ix*
Meet the Author		*xi*
Preface		*xiii*
	Exploring Active Meaning-Making	1
	Leading Effective Science Curriculum-Based Learning	3
	Addressing Contemporary Standards	5
	Safety in Science	7
	What This Guidebook Is About	8
	Intended Audience	9
	What This Guidebook Is Not About	10
Module 1	**What Does *Explore-Before-Explain* Look Like in Practice?**	13
	Thermal Energy Model Lesson	15
	Vignette: Exploring Ice Cold Lemonade	16
	Unpacking the Thermal Energy Model Lesson	24
	Thermal Energy Standards Connections and Progressions	27
Module 2	**Rethinking Learning**	30
	Modern Learning	31
	Call to Action	39
Module 3	**Exploring Sinking and Floating**	42
	Vignette: Exploring "Sinking and Floating"	44
	Unpacking the Exploring Sinking and Floating Model Lesson	52
	Exploring "Sinking and Floating" Standards Connections and Progressions	55
Module 4	**How Can We Make Science Learning Remarkable?**	57
Module 5	**Planning Engaging and Effective Science Lessons**	61
	Step 1: Plan "Backward" (Constructing Claims)	63

	Step 2: Activating Students' Ideas: Creating a Need-to-Know Situation	66
	Step 3: Enhance Understanding (Connecting Claims to Scientific Principles)	68
	Step 4: Promoting Reflection on Learning	71
	Step 5: *Explore-Before-Explain*	73
	Conclusions	75
Module 6	**Exploring Fall Time**	**78**
	Vignette: Exploring "Fall Time"	80
	Unpacking the Helicopter Fly Time Model Lesson	88
	Helicopter Fly Time Standards Connections and Progressions	90
Module 7	**Applying Active Meaning-Making and *Explore-Before-Explain* to Practice**	**92**
	Step 1: Plan "Backward" (Constructing Claims) in Practice	93
	Step 2: Engaging Students' Ideas in Practice	106
	Step 3: Enhance Understanding (Connecting Claims to Scientific Principles) in Practice	109
	Step 4: Promoting Reflection on Learning in Practice	112
Module 8	**A New Mindset to Teaching**	**117**
Module 9	**Leading Change**	**121**
	An *Explore-Before-Explain* Mindset to Teaching	123
	Conclusions	**126**
	Index	*130*

Foreword

In today's science education landscape, a key to ensuring that all students can learn, appreciate, and enjoy science lies in supporting effective instructional approaches through collaborative professional learning and leadership. This book emerges as a beacon for instructional leaders, offering not only theoretical insights but also practical guidance rooted in modern standards and research. It supports implementation of the *Explore-Before-Explain* approach that teaching science should be an active process where students don't simply absorb information but take charge of their own learning—surfacing their initial ideas, questioning, gathering evidence, analyzing data, and making sense of new ideas in a meaningful way. By championing collective leadership and a commitment to evidence-based teaching, this book provides the support and guidance needed to ensure that every teacher and student can experience the richness and relevance of science.

One of the book's greatest strengths is its approach to connecting research to practice. By grounding instructional sequence in well-established research, it ensures that the instructional methods described are not just theoretical ideas, but actionable approaches that support student learning. The clear principles laid out by Patrick Brown provide instructional leaders with the conceptual understanding needed to craft lessons that are purposeful, effective, and student-centered. With a focus on the *Explore-Before-Explain* instructional sequence, this book provides leaders with model lessons and practical tools to confidently implement approaches through which students grapple with new concepts and phenomena and build their own understanding before formal explanations are introduced.

Reflection is a cornerstone of professional learning encouraged throughout this book. As the reader pauses and assesses their own practice, it fosters a culture of continuous improvement, where instructional leaders build confidence to adapt and refine their instructional strategies to better meet the needs of their students. This reflective process is complemented by the book's strong emphasis on collaboration among teachers, recognizing that collective efficacy—when teachers work together toward a common goal—can have a profound impact on both teaching and learning.

This is more than just a guide to teaching science—it is a comprehensive resource for building a learning community in which teachers can thrive. Through active learning, collaboration, and research-driven practices, this book offers an instructional pathway to meaningful implementation of today's science standards. It recognizes that effective science teaching and learning requires not just strong teachers but also strong leaders who can cultivate a collaborative, research-driven, and reflective school culture.

As an advocate for understanding students' thinking and using their ideas as a starting point for instruction, I support this book's approach to listening to students, guiding them from their initial conceptions toward deeper scientific understanding, and valuing their thinking throughout the learning process. In 1968, educational psychologist David Ausubel famously said, "The most important single factor influencing learning is what the learner already knows. Ascertain this and teach accordingly." While many of us recognize the importance of starting with students' ideas, the "teach accordingly" remains the challenge. This book is the turn-to resource that will help teachers adapt teaching strategies accordingly. By doing so, it ensures that instruction is responsive, rooted in students' current understanding, and promotes deeper and more meaningful learning experiences.

Page Keeley, Author of *Uncovering Student Ideas in Science* and Past President of NSTA and NSELA

Acknowledgments

This resource is not just a guide—it is a testament to the leadership lessons that have shaped my journey, many of which stem from a coaching perspective of professional learning. I owe a deep debt of gratitude to my high school coaches: Coach Kern, Coach Jeffery, Coach Beckmann, and in memory of Coach Beard. The title "Coach" in my dedication signifies more than just a role; it represents my profound respect for these four individuals who have left an indelible mark on my life. Through their unwavering belief in me, I learned the value of persistence, the strength of a growth mindset, and the transformative power of tackling challenges with optimism. These lessons, forged on the cross-country course and in the wrestling room, have not only influenced my approach to education but have also been guiding principles in all aspects of my life. It is with heartfelt appreciation that I dedicate this book to them.

I would like to express my deepest gratitude to the incredible NSTA Press team: Emily Brady and Cathy Iammartino. Not only do they expertly manage the production and proof of concept, but they are true partners in this journey. Their attention to detail, creativity, and dedication mirrors my own commitment to writing, and together, we bring forward resources that help teachers inspire and equip students in science. Much like my role in crafting this book, they play an essential part in shaping its success, ensuring that the mission of better science education reaches the classrooms where it is needed most. Thank you for being valued collaborators every step of the way.

I would also like to extend my heartfelt thanks to Julia Dolinger at Routledge. Your visionary leadership and collaborative spirit have been instrumental in merging teams and aligning efforts to ensure this book reaches its full potential. Your ability to see the bigger picture and guide the press team toward making this work as impactful as possible has truly elevated the project. Thank you for your dedication and for making this journey a shared success.

I am deeply grateful for the opportunity to collaborate with Rodger Bybee and Jay McTighe, whose insights have helped anchor active meaning-making from a pedagogical perspective.

My heartfelt thanks go to Page Keeley for being an inspiring thought partner on *Explore-Before-Explain* teaching.

To all the teachers and leaders who have participated in my professional learning experiences over the years, thank you for your valuable insights and dedication to making *Explore-Before-Explain* a reality for all students. Your contributions have truly shaped this work.

Meet the Author

You might wonder how my experiences have shaped this book's content. My journey has been diverse and enriching, spanning roles as a classroom teacher, educator in various prospective teacher development programs, leader in professional learning development and implementation, and currently, as the Executive Director of STEAM and CTE for the Fort Zumwalt School District in St. Charles, MO.

My passion for ensuring students are success-ready in K-12 education and beyond has driven my work. In my role as Executive Director, I collaborate closely with teachers to develop curricula that equip students with the knowledge and skills they need for today's and tomorrow's questions and problems, and a globally competitive workforce. Together, we approach curriculum development and instructional design from a Standards-minded, research-based perspective, always with students' needs at the forefront.

My previous roles as a science coordinator and classroom teacher have deeply influenced my approach to developing STEAM and CTE programs. I have focused on essential lesson components, bundling them to create impactful curricula. My extensive writing on sequencing science lessons through my National Science Teaching Association (NSTA) books series *Instructional Sequence Matters* reflects my dedication to translating effective teaching practices into broader educational impacts. My dissertation was an investigation of the factors that facilitate and constrain teacher learning and grounds much of my views about how to best prepare teachers to teach in ways that align with modern research. My work comes from both research and practitioner mindset. I have published articles in research journals such as *Science Education*, the *International Journal of Science Education*, and the *Journal of Science Teacher Education*, and articles for teachers in *Science and Children*, *Science Scope*, and *The Science Teacher*.

Working as a classroom teacher, developing curriculum, and creating programs for students has taught me about the challenges many educators face. I have learned in my transition to a leadership role that our expectations of teachers are higher than ever and that we must balance how we honor educators' talents and gifts with innovative ways to better prepare students for the real world. My hope is that my work enables students to leave school with the emerging skills and knowledge to make informed decisions in their lives, be competitive in postsecondary education, and be the best candidates for professions they aspire to through the more intentional preparation of teachers.

Pat Brown

Preface

At its core, this guidebook is founded on the inspiring belief that teachers are pivotal in shaping student learning experiences. It's all about helping teachers develop a strong sense of their abilities to nurture student learning and success. Imagine teachers who grasp the latest research on effective teaching and learning and understand how to apply these principles in the classroom. They become equipped with a solid foundation, influencing their beliefs about creating the best possible teaching and learning environments. However, it does not stop there. Working together as a team of teachers can remarkably impact student motivation and learning. It is a ripple effect that can be felt at every level, from grade-specific classrooms to entire school districts and nationwide.

I am passionate about promoting active learning, where students do not just passively receive information but actively seek to understand, analyze, and make sense of new ideas. It is about connecting the dots between further information and what they already know while reflecting on the factors contributing to their newfound knowledge. Moreover, here is the exciting part: this approach works for all students, regardless of their backgrounds or abilities. As a result, I illustrate the process and research through vibrant and vivid examples I have used with students and teachers alike. However, the examples through the model lessons are only part of the change process.

The research chapters are designed to provide a holistic understanding of learners from the early years of kindergarten to the final stages of twelfth grade. Research highlights that every student possesses incredible intellectual abilities and naturally uses patterns and cause-and-effect relationships to explore the world around them. Driving change takes more than participating in a model lesson and studying the research alone. This is where clear instructional design principles and district examples linked to research come in. The instructional design principles are a way of thinking more purposefully and carefully about the nature of *any* design that has students' active meaning-making as the goal. Because active meaning-making is a phrase used throughout this guidebook, it is worth describing up front. Active meaning-making refers to the process through which learners construct and derive meaning from their classroom experiences by engaging with authentic tasks and focusing on understanding rather than memorization (Wiggins and McTighe, 2005). Educational standards provide a foundation, but the real magic happens when teachers craft activities that truly engage students in active meaning-making using the essential elements offered in this guidebook.

The mindset chapter is included to help you (and your teams) reflect on your educational memories, the ever-evolving nature of knowledge, and the influential role that beliefs play in your teaching. Instead of viewing mindsets as fixed or growth-oriented, I provide practical examples and planning considerations rooted in research. The mindset chapter comes later in the book so teachers and teams can see and reflect on examples, learn about contemporary research, and have explicit ideas about instructional design principles. These three elements solidify your beliefs (your mindset) about instructional design. A substantial body of science education research supports the sequence of topics in this book to promote the development of teacher mindsets. Several studies (see Gess-Newsome, 2015; Gess-Newsome & Lederman, 2001), and some that I have been involved with (see Brown et al., 2013; Friedrichsen et al., 2009), show that a teacher's mindset (termed "orientations" in the science education literature) is well formed from various influences and impacts how they design and deliver instruction as well as how they perceive the goals for education. To be clear, this book advocates an *Explore-Before-Explain* mindset because it seamlessly aligns the essential elements of active meaning-making, modern science education Standards, and emerging research while using teachers' experiences and assets to develop knowledge and practice from a more pedagogical perspective. Acknowledging mindsets empowers all teachers to take meaningful steps toward professional growth within their abilities and skill sets. The unique blend of your beliefs, your understanding of research, and clear examples of effective practice and instructional design are essential in building your confidence as a teacher. Each factor must align consistently to establish high self-efficacy related to contemporary research.

While we can make a significant impact individually, the power of collaboration cannot be overstated. This book aims to do more than change a single teacher's mindset and resulting practices; it is all about leadership that instills change on a broad level. Teacher collective efficacy is the shared and enthusiastic belief among groups of teachers that they can profoundly impact student motivation, learning, and achievement (Hattie, 2008). Research strongly supports this idea. Collective teacher efficacy ranks as the most influential factor in student learning and growth (see Side Bar: Teacher Collective Efficacy). This guidebook's primary goal is to foster teacher collective efficacy, empowering educators to be a driving force for modern educational goals so all students are well prepared for success in grades K-12 and beyond.

> **Side Bar. *Teacher Collective Efficacy***
>
> Teacher collective efficacy is the shared belief among groups of teachers that they can profoundly impact student motivation, learning, and achievement (Hattie, 2023). By collectively believing that we can make a difference through our teaching, we can help all students, regardless of background, be high achievers. Hattie's (2008) landmark review that analyzed over 800 meta-analyses of research ranked collective teacher efficacy as the most influential factor influencing student learning and growth in a school year.

Reaching Our Goals

I encourage you to consider two essential questions when considering the guidebook's goals and holding yourself or your teams accountable for professional learning: "Accountable for what?" and "According to what measures?"

For "accountable for what," I have outlined specific professional learning targets for teacher leaders, including active participation and reflection on engaging active meaning-making lessons tied to standards, a deeper understanding of research-based learning, and the ability to connect key features with contemporary research and standards (see Side Bar: Accountable for What?).

> **Side Bar. *Accountable for What?***
>
> ★ Actively participate and reflect on vivid examples of learning by active meaning-making through *Explore-Before-Explain* lessons tied to Standards
> ★ Understand learning from a more research-based perspective and
> ★ Connect key features of learning and *Explore-Before-Explain* to contemporary research and Standards

As for "according to what measures," it is all about assessment. I invite you to use essential features of active meaning-making and *Explore-Before-Explain* techniques to inspire changes in your teaching practices. Document how your instructional approaches evolve due to the activities outlined in this guidebook.

I hope this guidebook inspires you to become an even more effective, enthusiastic, and impactful educator!

References

Brown, P., Friedrichsen, P., & Abell, S. (2013). The development of prospective biology Teacher's PCK. *The Journal of Science Teacher Education*, 24(1), 133–155.

Friedrichsen, P., Abell, S., Pareja, E., Brown, P., Lankford, D., & Volkmann, M. (2009). Does teaching experience matter? Examining biology teachers' prior knowledge for teaching in an alternative certification program. *Journal of Research in Science Teaching*, 46(4), 357–383.

Gess-Newsome, J. (2015). A model of teacher professional knowledge and skill including PCK. In A. Berry, P. Friedrichsen, & J. Loughran (Eds.), *Re-examining pedagogical content knowledge in science education* (pp. 28–42). Routledge. https://doi.org/10.4324/9781315735665

Gess-Newsome, J., & Lederman, N. (2001). *Examining pedagogical content knowledge*. Kluwer Academic.

Hattie, J. (2008). *Visible learning*. Routledge.

Hattie, J. (2023). *Visible learning: The sequel: A synthesis of over 2,100 meta-analyses relating to achievement* (1st ed.). Routledge. https://doi.org/10.4324/9781003380542

Wiggins, G., & McTighe, J. (2005). *Understanding by design* (Expanded 2nd ed.). ASCD.

Exploring Active Meaning-Making

We know from the cognitive sciences that there is remarkable consistency between how people learn best across ages. To deepen teachers' professional practice by creating meaning and understanding from complex information, it is essential that they gather and analyze information, identify patterns, and construct mental models that provide coherence and meaning. Active meaning-making often requires integrating different perspectives and revising one's understandings and beliefs based on new insights through reflection on practice and developing knowledge. Rather than tell what active meaning-making is and why it is so essential for learners, followed by classroom examples, this book is designed to be an immersive *Explore-Before-Explain* process for readers and leaders who wish to use it to develop professional learning experiences and drive professional learning communities' (PLCs) work. Following this format, readers will explore active meaning-making in practice before explaining the key features and conduct critical inquiry by investigating extensive research support (critical inquiry is a key component of effective PLC work; Dufour & Eaker, 1998). I invite readers to try the model lessons in professional learning and through PLCs as they read (with fellow teachers acting as proxies for students) and reflect on them to provide students with motivated and coherent learning experiences.

The model lessons were chosen because they can be completed quickly (30–60 minutes) and acknowledges there are only so many collaborative hours in a teacher team's schedule. Each model lesson uses different reflection questions (Idea Catchers) so teachers can make sense of the research and planning considerations necessary to promote active meaning-making. In addition, each model lesson includes stopping points for elementary, "just for elementary," and "just for secondary." These departure points help to illustrate where the lesson is grade-span appropriate according to the frameworks and offers all teachers—elementary and secondary alike—with an understanding of learning progressions and vertical alignment. Teachers may wonder how a lesson can be used across the K-12 spectrum. For teams with considerable vertical alignment, the elementary portions of lessons provide valuable insights into secondary students long-lasting conceptual understanding. Secondary teachers do not worry because the preassessments that work at elementary can be used with older students to set up more advanced learning and dictate how long you spend on lesson activities.

This guidebook uses an *Explore-Before-Explain* instructional sequence to facilitate the design of active meaning-making lessons (see Brown & Keeley, 2023). *Explore-Before-Explain* teaching aligns with cognitive science research and emphasizes that a student's construction of knowledge creates a framework for developing understanding. The hallmark of *Explore-Before-Explain* is sequencing instruction so students build their experiences and knowledge by using data and evidence before developing explanations. Said differently, *Explore-Before-Explain* lessons ensure that a student's conceptual understanding is constructed primarily on evidence-based experiences.

In addition, promoting visible learning is highlighted throughout the active meaning-making lessons and instructional design processes. Students should have ample opportunities to reflect on their learning and assess how far they have progressed intellectually (e.g., promoting metacognition), which has a substantial impact on learning (Bransford et al., 2000). Formative assessment is integral throughout, with teacher-student discourse centered on prompting them to ponder patterns, cause-and-effect relationships, and the development of their thinking rather than validating ideas. When people actively contemplate their evolving understanding, they significantly contribute to the active meaning-making process. John Hattie's landmark review in 2023, which analyzed over 800 research meta-analyses, ranks "visible learning" (i.e., students thinking about their understanding) as the second most influential factor in achievement. As students assess which experiences aid their development of a more sophisticated understanding, they gain the skills to become more self-directed learners.

While this guidebook is meant to be written in an *Explore-Before-Explain* format, covering a few additional topics is needed to set the stage before delving into what active meaning-making looks like in practice.

Leading Effective Science Curriculum-Based Learning

Leaders are expected to be transformational and catalytic. They are tasked with creating solutions based on problems, questions, and needs and develop processes that work for their schools, districts, and regions. Due to the variety of expectations placed on many leaders, their role is multifaceted and complex. One end of a leader's work is anchored in developing the knowledge and skills of their teachers, and on the other end, the need to ensure students are success-ready in K-12 and beyond.

Richard Elmore (2009) proposes that to improve student learning at a large scale we must focus teacher leadership on the instructional core. The instructional core includes three critical elements: (1) the need to reform based on the Next Generation Science Standards (NGSS) and associated new state standards; (2) develop instructional programming, i.e., a curriculum, that presents the innovations recommended in the new state standards; and (3) provide the professional learning that teachers will need for effectively teaching standards-based curricula. These three instructional elements can be thought of as analogous to a three-legged stool. Each leg must be designed and aligned with incredible precision to maintain balance.

While this book is a standards-based reference, the standards are already set (Instructional Core Element 1). The main purpose of this book is less about educating leaders on the standards and more about the innovations

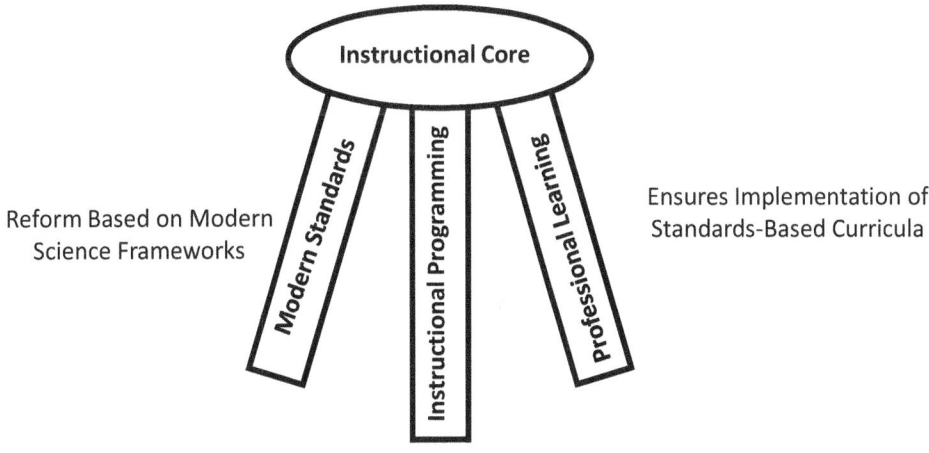

Figure 1.1 The three "legs" needed to improving learning at a large scale

required to make them a reality for students through more purposeful and meaningful instructional design that highlights active meaning-making practices (Instructional Core Element 2). Leaders should be well versed in the key features of a coherent, content-rich curriculum that offers a schedule of "what to teach and when" and is inspired by how students learn best.

High-quality professional learning is about creating an experience that is as engaging as it is enlightening and attempts to address both ends of the spectrum of a leader's goal (e.g., improve teacher practice and student learning). Leaders can look to the well-established research base on the characteristics of effective curriculum-based professional learning that emphasizes the following key features and is the infrastructure of the chapter sequence provided for leaders (Darling-Hammond et al., 2017; Short & Hirsch, 2022) (Instructional Core Element 3):

- **Content-focused and standards-aligned**: Deepens educators' understanding of what to teach and how to teach it within the context of the Standards, local curriculum, and high-quality instructional resources
- **Equity-focused**: Empowers educators to captivate every student, tailoring engaging tasks to diverse needs and abilities
- **Considerate of adult learners**: Addresses expressed and unexpressed expectations and motivations while attending to mindsets, builds on participants' prior knowledge and experience, and invites them to connect learning to meaningful goals and immediate valuable actions
- **Learner-centric**: Inquiry-based, interactive, and collaborative. Involves expert models and practice as educators participate in lessons as learners, plan, internalize, rehearse, observe, and reflect with colleagues who teach in the same content area and use the same curriculum
- **Provides coaching and expert support**: Offers expertise about curriculum, adopts high-quality instructional resources and evidence-based practices, focuses directly on educators' and students' individual needs
- **Offers feedback and reflection**: Provides job-embedded time for educators to think about intentionally, receive input, and refine practice; educators need adequate time to learn, rehearse, implement, and reflect upon new strategies that facilitate refinements in practice over time

Each chapter draws on one or more essential features of high-quality professional learning, and the entire guidebook incorporates all elements.

Sometimes, teachers might use a single model lesson with their team, acting as substitutions for students. Other times, teachers will sharpen their understandings based on emerging research to develop instructional practices from a more pedagogical perspective. In the end, teachers will be tasked with designing active meaning-making lessons and eliciting feedback from their colleagues. In each case, reflection is paramount and aimed at helping teachers play an active role in their learning.

The crucial components of effective professional learning and the strategies provided to engage teachers in developing deep conceptual understanding must be done regularly and obsessively in our schools and district training and professional learning communities (PLCs), as well as in graduate and undergraduate coursework. If we do these things, we can expect to profoundly impact student motivation, learning, and achievement by building teachers' collective efficacy through more purposeful teacher leadership (Hattie, 2023).

Addressing Contemporary Standards

This guidebook is a Standards-based reference. *A Framework for K–12 Science Education* (NRC, 2012) and the *Next Generation Science Standards* (*NGSS*) (NGSS Lead States, 2013), which are based on the *Framework*, aim to shift instruction so that students can take a more active role in learning. "*Standards*" is used throughout this guidebook as an abbreviation for the *NGSS* which is based on the *Frameworks*. I show how *Explore-Before-Explain* supports the vision of teaching and learning described in contemporary science Standards. While this guidebook is standards-based reference, the standards are already set. Understanding of standards and unpacking them in professional learning is a common activity; it often falls short of teachers developing deeper "know how" and "how to" knowledge for putting modern reforms in practice (Wiggins & McTighe, 2011). Alas, as we all know from experience, what seems like a good idea—focusing professional learning on teaching educators about standards—has a common unintended consequence of teachers not developing the abilities to put the innovations required of Standards into practice. Thus, the purpose is to make the standards come alive through active teacher participation in model lessons, exploring research, and using practices associated with effective curriculum design and professional learning. What follows is a brief description of the architecture of modern science education goals for student learning.

The *Framework* advocates intertwining disciplinary core ideas (DCIs), science and engineering practices (SEPs), and crosscutting concepts (CCCs) to

Table 1.1 A summary of the three dimensions of the NGSS

Science and Engineering Practices	Crosscutting Concepts
1. Asking questions and defining problems 2. Developing and using models 3. Planning and carrying out investigations 4. Analyzing and interpreting data 5. Using mathematics and computational thinking 6. Constructing explanations and designing solutions 7. Engaging in argument from evidence 8. Obtaining, evaluating, and communicating information	1. Patterns 2. Cause and effect 3. Scale, proportion, and quantity 4. Systems and system models 5. Energy and matter 6. Structure and function 7. Stability and change
Disciplinary Core Ideas	
Physical Science PS 1: Matter and its interactions PS 2: Motion and stability: Forces and interactions PS 3: Energy PS 4: Waves and their applications in technologies for information transfer **Life Sciences** LS 1: From molecules to organisms: Structures and processes LS 2: Ecosystems: Interactions, energy, and dynamics LS 3: Heredity: Inheritance and variation of traits LS 4: Biological evaluation: Unity and diversity	**Earth and Space Sciences** ESS 1: Earth's place in the universe ESS 2: Earth's systems ESS 3: Earth and human activity **Engineering, Technology, and the Application of Science** ETS 1: Engineering design ETS 2: Links between engineering, technology, science, and society

form opportunities for what is called three-dimensional teaching and learning. While these dimensions are critical, they support students' understanding.

The three dimensions have been analogous to threads woven or braided into a rope, creating a more extensive and substantial form (NGSS Lead States, 2013). The dimensions can be combined differently in curriculum and instruction and do not stand alone. In other words, when students learn content, DCIs are combined with SEPs. SEPs are not taught alone as separate process skills; they are always connected with a DCI. Moreover, a third dimension, the CCCs, helps to unify students' understanding across disciplinary boundaries. Several states have created new standards based on the *Framework* and have developed three-dimensional standards similar to the *NGSS* performance expectations for assessment. Becoming familiar with the three dimensions of the *Framework* can help determine the shifts needed to use active meaning-making in an *Explore-Before-Explain* instructional sequence. Each dimension is briefly described in the following section.

Science and Engineering Practices: The SEPs require students to mesh skills that scientists and engineers use, such as questioning, developing and using models, investigation, explanation construction, and problem-solving, to develop new content ideas (Bybee, 2012). The *Framework* identifies eight SEPs essential in a K-12 science and engineering curriculum and is taught in the context of students' learning content (NRC, 2012) (see Table 1.1). These eight practices embody the multifaceted, overlapping processes scientists use to develop and share knowledge about the natural world. As students use these practices in the *Explore-Before-Explain* sequence of instruction, they generate knowledge and come to understand how knowledge is generated in science.

Crosscutting Concepts: The CCCs provide an organizational framework for helping students connect knowledge from different science disciplines. They aid students in gaining a deeper understanding because of their explanatory power, connecting knowledge from disciplines to a coherent and scientifically based view of the world (NRC, 2012). The *Framework* identifies seven CCCs spanning disciplinary boundaries (Duschl, 2012) (see Table 1.1). These CCCs help students focus their thinking on particular aspects of natural occurrences.

Disciplinary Core Ideas: The DCIs include statements of scientific ideas used to understand and explain natural and human-designed events around four core areas: physical sciences, life science, Earth and space sciences, and engineering technology and the application of science. Within each core area are DCIs, which describe the content central to each discipline and necessary for students to develop proficiency (see Table 1.1).

Safety in Science

This book advocates for teachers to try to model lessons in teacher teams and reflect on crucial active meaning-making elements. In addition, a goal of this book is that teachers use active meaning-making practices to transform their science lessons and curriculum. It is essential to implement safety practices within the context of science investigations. When you keep safety at the forefront of your mind as a teacher, you avoid many potential issues with the lesson while protecting your students. Teachers should be aware of and support any school or district safety policies, legal safety standards, and better professional practices that are in place, and investigations should apply those safety protocols that align with the work being conducted in the lesson.

Safety practices encompass things considered in the typical science classroom (e.g., wearing safety goggles or safety glasses with side shields, vinyl gloves, and non-latex aprons as appropriate). At the same time, other focus areas, such as engineering, require students to demonstrate how to use the equipment before allowing their use. Science investigations should always be supervised, and safety procedures should be reviewed before initiating hands-on activities or demonstrations. Each lesson within this module includes teacher guidelines for applicable safety procedures that should be followed. For each investigation, teachers should remind student teams precisely what safety procedures they should follow.

Information about classroom science safety, including a safety checklist for science classrooms, general lab safety recommendations, and other science safety resources, is available at the Council of State Science Supervisors website at *www.csss-science.org/safety.shtml*. The National Science Teaching Association (NSTA) lists science rules and regulations, including standard operating procedures for lab safety and a safety acknowledgment form for students and parents or guardians to sign. In addition, NSTA's Safety in the Science Classroom web page (*www.nsta.org/safety*) has numerous links to safety resources, including safety papers written by the NSTA Safety Advisory Board.

Disclaimer: The safety precautions of each activity are based in part on use of the recommended materials and instructions, legal safety standards, and better professional practices. The selection of alternative materials or procedures for these activities may jeopardize the level of safety and, therefore, is at the user's own risk. Additional information regarding safety procedures can be found on other NSTA sites, including the NSTA Safety Portal: Safety in the Science Classroom (*www.nsta.org/safety*).

What This Guidebook Is About

This guidebook is about good teaching and learning at the lesson level. This resource is about spending more time developing students' conceptual understanding by investigating natural occurrences. The approach may feel new and different for teachers and students alike but focuses on putting emerging research into practice through "active meaning-making" arranged in an *Explore-Before-Explain* format. To accomplish these goals, this guidebook further develops ideas presented in my *Instructional Sequence Matters* series (see *www.nsta.org/book-series/instructional-sequence-matters*) and *Activating Student Ideas: Linking Formative Assessments to Instructional Sequence* (Brown & Keeley, 2023). The research is updated with emerging work on the essential elements needed for active meaning-making at all grade spans (see Brown & Bybee, 2024; Brown et al., 2023a, 2023b; Brown & Bybee, 2023a, 2023b; Brown et al., 2023).

This guidebook also promotes learning for all students. Students come to us with varied background experiences and ideas. In this regard, I have drawn on students' ideas research from my collaborations with Page Keeley (see Brown & Keeley, In Press; Brown & Keeley, 2023). In our planning, we must use students' backgrounds as assets for future learning. Promoting active meaning-making allows teachers to provide classroom experiences that benefit all students by becoming familiar with the research on learners and learning tied to critical elements of instructional design. By working together, students can construct scientific ideas through productive discourse that focuses on data and evidence from firsthand experiences. In collaborative teaching teams, individuals can share their interpretations to collectively make sense of what good teaching looks like in modern classrooms.

Finally, while this book does not offer an entire curriculum, the lesson level perspective offers teams the ability to think more purposefully about how they bundle together individual lessons into more cohesive lessons and units, addressing multiple topics and standards. For many teachers, having a clear understanding of how to engage students in learning by doing and constructing knowledge through *Explore-Before-Explain* allows them to hook on related topics to further elevate student learning.

Intended Audience

This guidebook is for teachers, new and veteran educators who want to be leaders, including building administrators, through improved instructional practices across the K-16 education continuum. I have worked with teachers who use loads of hands-on activities. While their students enjoy science, their lessons need to promote deep conceptual meaning-making. On the other end of the spectrum, I have worked with teachers who mostly deliver the content-rich portions of science through lectures and use hands-on to confirm for students what they have been told. While lectures often include interesting stories that are often student favorites, many students leave their classes not knowing from the lecture, "What science is important here," and "How is this knowledge relevant to my future life?" The approach in this guidebook is not an either-or proposition—hands-on versus lecturing. Instead, educators can make simple shifts in instructional practices and focus on the most salient features of how people learn for their lesson design. Students will confront limits in their understanding, leading to where to go next. This book honors both hands-on and delivery modes of teaching, realizing there is a right time and place to introduce ideas and it is after students have experienced learning by doing.

The guidebook is meant to take teachers and leaders through a practice-oriented approach and then use vivid examples to develop professional knowledge from a more research-based perspective. Teachers gain a powerful strategy for leveraging the research and hands-on, minds-on activities to promote more robust and equitable learning environments. Science education leaders gain a resource that can be used to develop professional learning experiences and drive professional learning community efforts aimed at sustained improvement. I want science leaders to use the guidebook with teacher teams to create higher levels of collective efficacy.

What This Guidebook Is Not About

This guidebook is not about providing expansive model lessons—those are available in my *Instructional Sequence Matters* series and *Activating Student Ideas: Linking Formative Assessments to Instructional Sequence* (Brown & Keeley, 2023) books and the large corpus of examples in the NSTA journals. Teachers and leaders can easily use other examples from my books and NSTA journals using a format similar to this guidebook to tailor professional learning to meet their needs.

This guidebook is also not a prescriptive program but rather considers key elements of instructional design in light of research. As such, many opportunities exist to highlight STEM applications, using problem-based or project-based learning, and emphasizing inquiry. The approaches depend on the content and activities planned for students and seamlessly integrate with the active meaning-making features discussed. This guidebook is also one of many ways to teach in the vision of the Standards. The guidebook could be a keystone to developing more elaborate approaches. A primary goal of this guidebook is that teachers will use the ideas to build lessons and curricula that meet their students' needs. I offer some guidance on content examples from my past experiences with students. However, the focus is that our best lessons should include the critical elements of active meaning-making when designing an *Explore-Before-Explain* sequence.

Throughout the guidebook, I ask teachers and teams to explore and reflect on key ideas. Readers, brace yourself and realize that the more thoughtful and attentive teachers are about the ideas in the guidebook, the more you will learn. Use the reflection prompts to think about the critical features of active meaning-making. The reflection prompts are designed to place teachers as learners and experience the environment, which is true of students who *Explore-Before-Explain* science.

References

Bransford, J., Brown, A., & Cocking, R. (2000). *How people learn: Brain, mind, experience, and school.* National Academies Press.

Brown, P., & Bybee, R. (2023a). Promoting sensemaking: An impactful instructional sequence for teaching elementary students whether objects are heavy or light for their science. *Science and Children, 60*(4), 30–33.

Brown, P., & Bybee, R. (2023b). Promoting sensemaking through an impactful instructional sequence. *The Science Teacher, 90*(6), 22–27.

Brown, P., & Bybee, R. (2024). Promoting sensemaking through an impactful instructional sequence. *Science Scope, 90*(6), 22–27.

Brown, P., Fries-Gaither, J., & Renfrew, K. (2023). Growing students' meaning making. *Science and Children, 60*(7), 58–63.

Brown, P., & Keeley, P. (In Press). *Activating students' ideas: Linking formative assessment probes to instructional sequences, grades 6–8.* NSTA Press and Routledge.

Brown, P., & Keeley, P. (2023). *Activating students' ideas: linking formative assessment probes to instructional sequence, grades K-5.* NSTA Press.

Brown, P., McTighe, J., & Bybee, R. (2023a). Promoting learning for all through explore-before-explain. *The Science Teacher, 90*(7), 24–27.

Brown, P., McTighe, J., & Bybee, R. (2023b). Leadership matters: Activating student learning through explore-before-explain. *Science and Children, 60*(6), 8–10.

Bybee, R. W. (2012). The practices of science: The influence of the National Research Council and the Next Generation Science Standards. *Science & Children, 49*(7), 10–16.

Darling-Hammond, L., Hyler, M. E., & Gardner, M. (2017). *Effective teacher professional development.* Learning Policy Institute.

DuFour, R., & Eaker, R. (1998). *Professional learning communities at work: Best practices for enhancing student achievement.* National Education Service.

Duschl, R. A. (2012). The crosscutting concepts: Strengthening science and engineering learning. *Science & Children, 50*(2), 34–39.

Elmore, R. (2009). The instructional core. In E. A. City, R. F. Elmore, S. E. Fiarman, & L. Teitel (Eds.), *Instructional rounds in education: A network approach to improving learning and teaching* (6th ed.). Harvard Education Press.

Hattie, J. (2023). *Visible learning: The sequel: A synthesis of over 2,100 meta-analyses relating to achievement* (1st ed.). Routledge. https://doi.org/10.4324/9781003380542

National Research Council. (2012). *A framework for K-12 science education: Practices, crosscutting concepts, and core ideas*. The National Academies Press. https://doi.org/10.17226/13165

NGSS Lead States. (2013). *Next Generation Science Standards: For states, by states*. National Academies Press. https://www.nextgenscience.org/next-generation-science-standards

Short, J. B., & Hirsch, S. (2022). *Transforming teaching through curriculum based professional learning: The elements*. Corwin.

Wiggins, J., & McTighe, J. (2011). *Understanding by design guide to creating high quality units*. ASCD.

Module 1

What Does *Explore-Before-Explain* Look Like in Practice?

Purpose: To investigate an *Explore-Before-Explain* instruction sequence to explain how essential learning elements can be arranged and highlighted to promote active meaning-making.

Desired Results:

Curriculum designers and teacher leaders will understand that:

- ★ Teachers benefit from engaging in a content-focused and standards-aligned experience around an expert model lesson so they better know what to teach and when
  - ○ Standards are not a curriculum by themselves; a curriculum works with classroom activities and their purposeful sequence to make standards come alive for students
  - ○ *Explore-Before-Explain* sequences of instruction naturally and seamlessly integrate the three dimensions of the *Framework*
- ★ Promoting reflection on practice creates more equitable experiences that tailor tasks to diverse teacher and student needs
- ★ Unpacking essential planning characteristics and elements of learning helps teachers think intentionally about instructional design, receive input, and offer feedback to advance the professional learning community's thinking.

Module Design Goals: In this module, you will perform an *Explore-Before-Explain* lesson that takes no more than 50 minutes to experience essential elements of learning arranged in an *Explore-Before-Explain* sequence. The end product will be a more sophisticated view of the importance of sequence of instruction and the implications for instructional design.

You should work on Module 1 if you are accustomed to traditional teaching methods emphasizing teacher explanations and unfamiliar with *Explore-Before-Explain* teaching. In addition, you should work on Module 1 if you are vetting the purchase of a pre-established curriculum to ensure it will meet students' needs and modern science education standards.

I invite you to *explore* a thermal and kinetic energy transfer model lesson with three video resources. The lesson study is arranged so educators can see how similar activities can be used at different grade spans (e.g., 3–5, 6–8, 9–12) to promote high levels of Standards-based learning (see Standards Connection and Progressions). In addition, the content of this particular model lesson differs from those to come in the guidebook due to the nature of the content, and I present adult ideas about the conceptual ideas to spark discussions about common misconceptions across ages. Consider other adult ideas to your teams' thoughts (or individual ideas).

This two- to three-day lesson was developed using the critical elements of active meaning-making arranged in an *Explore-Before-Explain* instructional sequence. The lesson includes how the activities play out in elementary and secondary settings. Elementary educators may wish to perform secondary education activities to understand the vertical learning progression. Secondary educators should perform the model lessons entirely, realizing they may need less time up front if they have a well-designed vertical curriculum.

Educators may wish to perform the lesson in professional learning groups to read, perform (or use video resources), and reflect on the lesson using the "idea catcher" questions below. In addition, reflection questions are embedded throughout the model lessons. The embedded reflection questions can be used in groups or individually and highlight the most salient aspects of instructional design that leverage high levels of active meaning-making.

Reflection Alert! Idea Catcher
- ★ *How are students' ideas activated at the beginning of the lesson?*
- ★ *How does the lesson target relevant conceptual science ideas?*
- ★ *How does student exploration lead to students' evidence-based claims?*
- ★ *How are assessments incorporated?*
- ★ *What teacher enhancements are needed to develop a deeper understanding?*
- ★ *How is Explore-Before-Explain being used?*

Thermal Energy Model Lesson

Materials Needed for the Lesson
Elementary and Secondary

- Ice-cold Lemonade Formative Assessment probe
- Hot water
- Cold water
- Food dye: Suggested dyes include red and blue food coloring
- Small travel-sized container like mouthwash (if available, a 50 ml Erlenmeyer flask)
- Medium-sized glass coffee mug (if available, a 500 ml beaker)
- Tongs to place Erlenmeyer beaker containing water (either cold or hot) in a beaker containing water (either cold or hot)
- Thermometer
- Indirectly vented chemical splash goggles
- Nitrile gloves
- Non-latex aprons

Secondary

- Mixing Water Formative Assessment probe (included)
- Thermometer
- 500 ml beaker (if not available, medium-sized glass coffee mug)
- Red and blue balloons

Safety Considerations

- Wear sanitized indirectly vented chemical splash goggles meeting ANSI/ISEA Z87.1 D3 standard, nitrile gloves, and non-latex apron during the setup, hands-on, and takedown segments of the activities.
- Secure loose clothing, wear closed-toe shoes, and tie back long hair.
- An eyewash station is required in case of a splash with hazardous chemicals or contact with physical hazards.
- Hard printed or electronic copies of Safety Data Sheets for all hazardous chemicals used are required, e.g., for food coloring.
- Immediately clean up any liquid spilled on the floor, so it does not become a slip/fall hazard.

- Follow procedures set up by the instructor for disposal of materials.
- Use caution when working with hot water! It can burn skin on direct contact.
- Use caution when working with glassware or plasticware. It can break and cut/puncture skin.
- Wash hands with soap and water or use hand sanitizer wipes immediately after completing this activity.
- Do not eat or drink anything used in this activity.
- Direct supervision will be required during all aspects of these activities.

Vignette: Exploring Ice Cold Lemonade

Students were asked to think about the saying and their thoughts and experiences, with an emphasis on connecting the phenomenon to real-world applications. For instance, they discussed how they think thermal energy transfer plays a role in processes like ocean currents and global warming. They were also encouraged to discuss why ice might melt faster in different environments. This linkage reinforced the broader relevance of the task. To situate and test students' ideas, they engaged in a hands-on exploration to allow them to make observations connected to thermal energy transfer. Students considered the "Ice Cold Lemonade" probe that asks students to think about how thermal energy transfers between ice and lemonade to address students' prior knowledge and situated learning around conceptual science ideas (Brown & Keeley, 2023; Keeley et al., 2007). Students were asked to think about the saying and their thoughts and experiences. Most students believed both hot and cold move simultaneously (59%). A few students (30%) thought it only went from hot to cold, and 11% believed it went from cold to hot. Once students had made their choices, they turned to a shoulder partner (two students total) and shared their thinking. Students based their thinking on the rule that when you touch something hot or cold, you feel those sensations (either hot or cold). For example, they all knew that holding a piece of ice made them feel cold, and over time, the ice would melt; however, their lived experiences did not provide evidence of the direction in which heat moves. While their lived experiences provided them with different ideas and rules, the driving conceptual ideas all students experienced invoked questions and curiosities that would be explored more deeply.

> **Side Bar. *Student Ideas Alert!***
>
> **These are the common student ideas identified in the research.**
> ★ Many researchers have found that children have difficulty understanding heat-related ideas. Harris (1981) and other researchers suggested that much of the confusion about heat comes from the words we use, and children think of heat as a substance that flows from one place to another. Cold is also considered an entity like heat, with many children thinking cold is the opposite of heat rather than being part of the same continuum (Driver et al., 1985, 1994).
> ★ Middle school students often do not explain the process of heating and cooling in terms of heat being transferred. When transfer ideas are involved, some students think cold is being transferred from a colder to a warmer object. Other students believe that both heat and cold are transferred at the same time (AAAS, 1993).
> ★ Middle and high school students do not always explain heat-exchange phenomena as interactions. For example, students may say that objects tend to cool down or release heat spontaneously without acknowledging that the object has come into contact with a cooler object or area (AAAS, 1993).

> **Side Bar. *Teacher Ideas Alert!***
>
> I have asked kindergarten through fifth-grade teachers whether they teach thermal energy and their ideas about the formative assessment probe. Many teachers who do not teach this topic have similar ideas and rules for their thinking as students. Figure 2.1 shows K-5 teachers' data about how they think thermal energy will transfer.

> **Reflection Alert! Would your teachers think similarly or differently?**

At this point, students were inquisitive about which direction heat moves (secondary students should think in terms of energy transfer). The opening wonderment scenario easily lends itself to similar testable situations and could produce the data for the class. So, the class began creating a scientific model to help explain and test predictions about the direction thermal energy transfers. Using an Erlenmeyer flask, a beaker, hot water, cold water, red food coloring, and blue food coloring, the class set out to answer which way thermal energy transfer. The class used two setups: (1) place hot water in the Erlenmeyer flask,

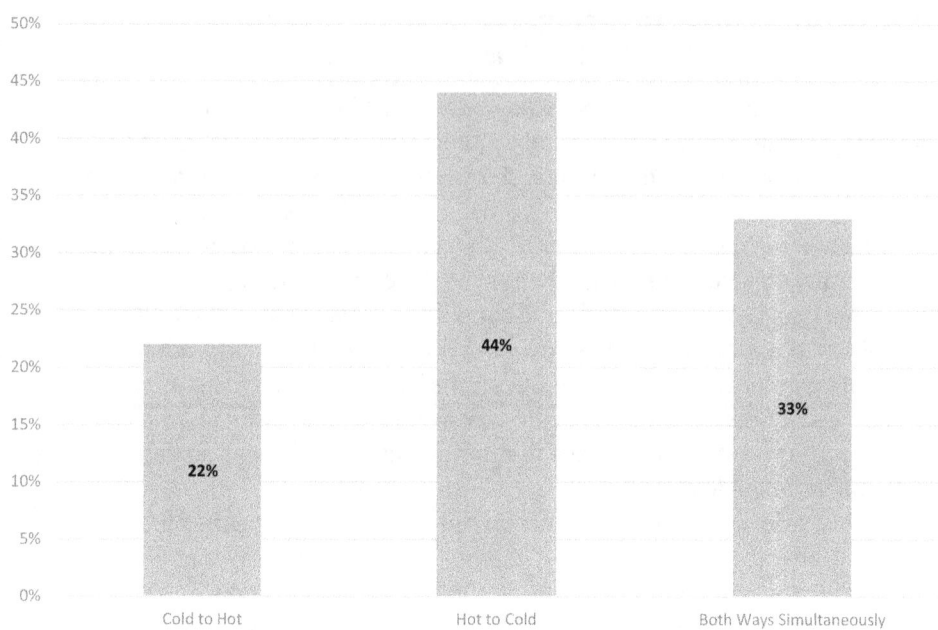

Figure 2.1 Adult preconceptions about the "Ice Cold Lemonade" probe

dye it red, and place it inside a beaker of cold water that was not dyed, and (2) place cold water in the Erlenmeyer flask, dye it blue, and place it inside a beaker of hot water that was not dyed. (As a note to teachers, the "hot water" was 65.5°C, and the cold water was 9.5°C. Teachers should wear protective goggles, rubber gloves, a heavy-duty rubber safety apron, and gloves to avoid burns from the "hot water.") Once students understood the experimental setup, they identified how the investigation would test how thermal energy transfers.

Students were asked to transfer their ideas from the "Ice Cold Lemonade" probe to the scientific model. Transferring their ideas to the model was to ensure they could identify the parts described in the probe and the model that would be used to explain and allow predictive value for future instances. The idea that scientific models must do two things—allow for explanations and predictions—was emphasized and made explicit to students as a way to use logical and critical thinking to know and understand their world through model-based reasoning.

Students had an intellectual stake in their learning. They shared their ideas for the formative assessment probe. In addition, they translated their ideas for the formative assessment probe into a scientific model. To say they wanted confirmation of their ideas was an understatement. The time spent up front eliciting ideas and thinking deeply about their rationales and rules would be made up for on the back end of instruction. Said differently, students' learning was activated, and they deeply desired to learn more about how thermal energy transfers.

> **Reflection Alert! Why are the pre-exploration activities that activate students' ideas important for learners?**

The demonstration allowed students to visualize heat and have a model of conceptual science ideas not readily observable first-hand. As a model, the demonstration allowed for predictive value and an explanation of science. To the students' surprise, heat moved in only one direction: from hot to cold. Students also had data from the model that heat did not transfer simultaneously from cold to hot or in both directions. (Video demonstration found here: *www.youtube.com/watch?v=xLA1EiXUCuM&disable_polymer=true*). Many students commented that their ideas had changed due to the demonstration. Even students who chose the correct answer developed a more sophisticated understanding because their conceptions were grounded in data-based experiences.

> **Reflection Alert! How does the demonstration provide data that serves as evidence for active meaning-making?**

The enhancements focused on helping students use scientific terminologies to explain their evidence-based claims and better reach the vision of the Standards. In pairs, students pretended that their fingertips were particles. One student's fingertips represented "hot," and the others acted as "cold" particles. Students acted out the demonstration using their fingertips to show the thermal energy transfer they witnessed. At this point, the term energy was introduced for the speed of fast or slow particles moving. Faster-moving particles had more energy than slower-moving particles. Energy can be moved from place to place by moving particles at different speeds.

> **Reflection Alert! How are new ideas bridged to students' firsthand experiences?**

Just for Elementary Students

Now that students had evidence-based experiences with thermal energy transfer, they revised their initial ideas discussed at the beginning of the lesson. Students changed their answers to provide a scientific explanation based on their experiences during the lesson. The demonstrations provided the qualitative evidence students needed to develop knowledge of the assessment probe. Students wrote a claim based on evidence statements to articulate their thinking. For example, one student wrote, "Heat energy from hot to cold but not cold to hot." They said the model "showed that only the warm water dyed red moved into the cold water. Cold water dyed blue did not move into the surrounding warm water."

Just for Secondary Students

Students had experienced the demonstration and were using the term "energy" to describe the motion of particles in the demonstration. They used a short section from their textbooks to introduce new terminologies related to energy transfer. Students read a short section of their textbook on convection, radiation, and conduction. Students applied these terms to their experiences with the demonstration to describe processes that happen together to explain the transfer of energy. Students drew the Erlenmeyer flask containing hot water moving into the cold water in the beaker and tabled the energy transfer by convection. At this point, I introduced "kinetic energy" so students could use scientific vocabulary to explain energy transfer at a molecular level. Next, students passed around a beaker of water. Then, they drew a hand touching a warm beaker of water and labeled it energy transfer by conduction. Finally, students considered feeling warm due to being in the sun. Students drew a picture of being warmed by the sun and labeled it energy transfer by radiation. In each of these instances, we used an integrated approach to teaching vocabulary where students connected their classroom and lived experiences to new terminologies (versus copying definitions of these terms).

> **Reflection Alert! How are new terms anchored on students' learning?**

The lesson continued by exploring the "Mixing Water" water formative assessment probe that asks students what the resulting temperature would be when two different water temperatures of equal volume are mixed (Brown & Keeley, 2023; Keeley et al., 2007). Students were also asked to think of a rule for their predictions. In addition, to make the concept of kinetic energy transfer more engaging and relatable, the class discussed real-world examples, such as how kinetic energy affects weather patterns through phenomena like land and sea breezes, and how kinetic energy transfer is managed in technologies like insulated drink containers or radiant heating systems. The lesson also incorporated familiar contexts, such as the cooling of beverages with ice cubes and strategies for maintaining comfortable indoor temperatures during seasonal changes. These discussions were further expanded to explore the implications of energy transfer in addressing global challenges, such as designing energy-efficient homes and mitigating urban heat islands.

Students used mathematical reasoning when formulating their rules. For example, more than half the students (57%) believed the correct answer to be C and reasoned that the resulting temperature would be a subtraction property. Others thought the answer would be B because of averaging the two temperatures (40%). A few students thought that the warmer temperature (50°C) might override the colder temperature (10°C) and result in 50°C water (3%). No students thought the combined temperatures would result from adding the two temperatures (answer E) or a mathematical equation that is not intuitive or easy to solve (answer A) (see Side Bar 2).

Student Ideas Alert!

★ Numerous studies have shown that few middle and high school students understand the molecular basis of heat transfer after instruction. Furthermore, difficulties in understanding remain even with instruction designed to explicitly address the difficulty of understanding heat transfer (AAAS, 1993).

★ Researchers have found that difficulties experienced by students in response to questions that ask them to predict the final temperature of a mixture of two quantities of water, given the initial temperature of the components, depending on the form in which the temperature problems are presented. Qualitative tasks in which the water is described as warm, cool, hot, or cold are easier than quantitative ones in which specific temperatures are given. The mixing of waters at different temperatures (e.g., hot and cold or 30°C and 80°C) is more complex than mixing water at the same temperature (e.g., warm and warm or 50°C and 50°C). It is not until around the age of 12 that most students can predict quantitatively what will happen in the type of problem posed in this probe (Erickson & Tiberghien, 1985).

★ Student responses to tasks similar to the one posed in this probe have been categorized according to the strategy used. Younger students (ages 8–9) prefer an addition strategy, whereas older students are more apt to use a subtraction strategy, which at least acknowledges that the final temperature lies somewhere in between. However, students ages 12–16 were as likely to use an addition or subtraction strategy as an averaging strategy (Erickson & Tiberghien, 1985).

> **Adult and Student Ideas Alert!**
>
> I have asked four different populations about their ideas for the mixing water formative assessment probes: grades 3–5 students, grades 6–8 students, adults in my elementary science methods courses, and adults in my secondary science methods courses (see Figure 2.2). Typically, adults in my elementary science methods courses have had one or two postsecondary science courses. Adults in my secondary science methods courses work towards undergraduate science degrees and teacher certification. Most of my secondary science methods students are biology majors. Figure 2.2 shows student and adult data about the mixing water temperature probe.

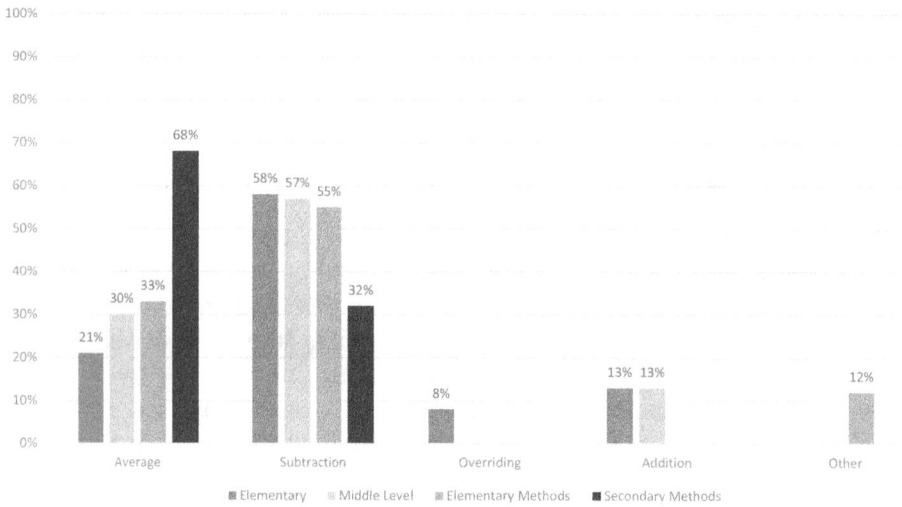

Figure 2.2 Adult and student preconceptions about the mixing water temperature probe

> **Reflection Alert! Would your students think similarly or differently?**

Once students discuss their thoughts about what the temperature will be when different water temperatures are mixed (subtraction, average, addition, no change, or other temperature), they set out to test these ideas with a partner (a video resource is here: www.youtube.com/watch?v=ejY-sgO0RUk). With safety goggles, protective aprons, and gloves, students select their own hot and cold temperatures but are not allowed to use boiling water (100°C). Students collect individual data and analyze class data for patterns and trends that can be used as evidence for active meaning-making. In addition, students compare their results to calculated values for the mixed temperature using subtraction and the average temperature (see Table 2.1).

Table 2.1 Class results for mixing water

Group	Cold (°C)	Hot (°C)	Mixed (°C)	Sub (°C)	Avg (°C)
1	19	84	49	65	51.5
2	12	84	48	72	48
3	15	85	53	70	50
4	15	80	47	65	47.5
5	15	88	51	73	51.5
6	7	92	49	85	49.5
7	27	82	52	55	54.5
8	15	88	51	73	51.5
9	20	90	55	70	55
10	25	90	56	65	57.5

Turning little data (i.e., single group) into big data (i.e., whole class) provides compelling evidence that the resulting product is the average temperature of the hot and cold water mixed. Even for students whose actual results are not 100% of the calculated average, they can see trends in the data that they are closer to a computed average than subtraction, as well as groups whose results are the same as the calculated average (e.g., in this class 50% of them resulted in the same value).

The class revisited acting out what happens when different temperatures of molecules with different kinetic energy come into contact with each other. Students faced a partner and acted out the relative molecular movements. They pretended that their fingertips were molecules. One partner's fingertips were hot molecules, and the others were cold molecules. Again, they used modeling to show how energy transfers from hot to cold until all molecules reach the average temperature. The molecular discussion aimed to develop a deeper conceptual understanding so that students modeled energy transfer between molecules and that matter is not transferred.

The last enhancement activity considered whether kinetic energy could influence the density of water. Students made predictions about whether they thought the density of hot and cold water was the same or different from water at room temperature. Hot water was placed in a red balloon. Cold water was placed in a blue balloon. Both balloons were placed in a 10-gallon fish tank filled with water at room temperature (a video demonstration is here: www.youtube.com/watch?v=qCIc5dhqPkg). From the demonstration, students had data that served as evidence that the temperature of water influences density. As a class, we discussed what was occurring at the molecular level that might explain the science concepts. Students drew on their past experiences to explain that differences influenced the molecules in each condition (e.g., hot and cold) in energy. In hot conditions, the molecules could spread out more than at room temperature. In cold water, the molecules were more closely packed together. As a result, the hot water was less dense than the room temperature and cold water.

Now that students had evidence-based experiences with energy transfer, they revised their initial ideas discussed at the beginning of the lesson. Students changed their answers to provide a scientific explanation based on their experiences during the lesson. The demonstrations provided the evidence students needed to develop knowledge of the assessment probe. Students could add scientific terminology and use "convection" as the mechanism of energy transfers through fluids.

For a summative assessment, students were asked to consider the following scenario: you have heard the saying, "Close the refrigerator door because you are letting the cold out," and someone else says in the winter, "Shut the front door, you are letting the heat out." Which description of the transfer of thermal and kinetic energy is more accurate? Students were asked to provide data as evidence from class to support their explanation.

Unpacking the Thermal Energy Model Lesson

These lesson elements targeted in the idea catcher and their sequence underscore a fundamental point about achieving deep and lasting learning—students need to actively "make meaning" to come to understand abstract science concepts (McTighe & Silver, 2020). In the next section, I unpack my impressions of the idea catcher reflection questions.

How Are Students' Ideas Activated at the Beginning of the Lesson?

In the model lesson, students' ideas are engaged at the onset of the unit using the ice-cold lemonade formative assessment probe. This initial engagement is crucial because it taps into students' prior knowledge and experiences, making the learning process more relevant and meaningful. Their ideas are further activated by translating their predictions into a scientific model that could be used to explain and predict thermal energy transfer. This process not only piques students' curiosity but also sets the stage for deeper inquiry. Using a scientific model engages students with science and engineering practices to "figure out" and explain conceptual ideas. For example, students might draw diagrams to represent how they think heat moves from the lemonade to the ice, prompting them to think critically and develop their initial hypotheses.

How Does the Lesson Target Relevant Conceptual Ideas?

All students have experience with cooling beverages using ice. This common experience is leveraged to make the abstract concept of thermal and kinetic energy transfer more tangible. Thus, the prompt sets goals for student learning that is motivated by curiosity—in other words, in which direction does thermal energy transfer? For secondary students, the mixing water formative assessment probe invokes questions about a common occurrence, such as why the temperature changes when different temperatures of water are mixed. This approach aligns with the constructivist theory of learning, which posits that learners construct new knowledge based on their existing understanding and experiences.

How Does Student Exploration Lead to Students' Evidence-based Claims?

For elementary and secondary students, the demonstration using different temperatures of dyed water in an Erlenmeyer flask and beaker provides data that supports their active meaning-making. This hands-on activity is essential because it allows students to observe the phenomenon directly, making the abstract concept more concrete. The demonstration is critical and serves as the focal point of the lesson because it provides students with a conceptual understanding of how "hot" and "cold" transfer in natural settings.

Secondary students have more sophisticated explorations that allow them to develop evidence-based standards-related claims. The mixing water exploration provides data indicating that when two different water temperatures are mixed, the resulting temperature is average. This activity helps students understand the principle of thermal equilibrium. In addition, secondary students elaborate on the ideas to consider how density and kinetic energy are related—an essential notion in physical, earth, and space science standards. For instance, they might explore how warm water rises and cold water sinks, leading to discussions about convection currents and their implications in natural phenomena such as ocean currents and weather patterns.

How Are Assessments Incorporated?

Assessments are used seamlessly throughout the lesson and serve many purposes. First, assessments like the ice-cold lemonade create relevancy and activate student thinking at the lesson's onset. This formative assessment helps gauge students' preconceptions and readiness for new learning. For secondary students, formative assessment probes are used to develop relevancy when further exploration is needed to address content in standards (e.g., mixing water). These probes prompt students to make predictions and explanations, fostering a mindset of inquiry and investigation. The conversations that occurred and revisiting initial ideas are also a formative assessment and promote students thinking about their developing understandings. Teacher-student discourse allows students to struggle productively and focuses on asking students to think about patterns and cause-and-effect relationships and developing thinking versus introducing content and validating students' ideas. This discourse is vital for developing critical thinking and reasoning skills. Finally, a summative assessment is included for secondary students to transfer their learning to new and different situations. This assessment could take the form of a project or presentation in which students apply their understanding to real-world contexts, demonstrating their ability to synthesize and extend their learning. The secondary summative assessment helps develop skills for students to take action for themselves and their families when encountering new and other challenges, such as understanding how home insulation works or why certain materials are better for keeping things warm or cool.

What Teacher Enhancements Are Needed to Develop a Deeper Understanding?

Teacher enhancements allow students to use scientifically accurate terminology in light of their firsthand experiences. The enhancement activities confront limits in student understanding and identify where the learning will go next. For elementary students in the probe, the term "energy" was introduced to label what students described as molecules' different speeds and temperatures. This introduction helps students begin to use precise scientific language to describe their observations.

For secondary students in the probe, the terms conduction, convection, radiation, and kinetic energy were introduced so students could use these labels to describe different ways kinetic energy transfers. These terms are critical for understanding more complex scientific concepts and for communicating their ideas effectively. Additional teacher enhancements might include using multimedia resources, such as videos or simulations, to visualize the molecular movement involved in kinetic energy transfer, thus deepening students' conceptual understanding.

How Is *Explore-Before-Explain* Being Used?

This two-to-three-day lesson on thermal and kinetic energy illustrates a critical point in curriculum planning and instructional sequencing: begin lessons by having students explore conceptual science ideas that are natural occurrences. This exploration phase is essential for activating prior knowledge and generating interest. Then, engage them in generating ideas, testing and collecting evidence, and making evidence-based claims. This approach encourages students to become active participants in their learning process. Then comes explanations by the teacher and elaborations to extend the learning. This sequencing ensures that explanations are more meaningful because they are connected to students' prior explorations and discoveries. This approach uses Science and Engineering Practices (one of the dimensions of the *Framework*) by engaging learners in "doing" science, not just learning facts delivered didactically. For instance, after exploring how different temperatures of water mix, students might engage in a discussion about the molecular basis of heat and/or thermal energy transfer, leading to a teacher-facilitated explanation that ties their observations to scientific principles.

An *Explore-Before-Explain* mindset approach that highlights active meaning-making acknowledges that the best learning environments activate students' ideas and offer them firsthand experiences, learning science by doing science. Once students have developed accurate understandings based on explorations, teachers enhance understanding through explanations at the right time for learners. This approach not only improves comprehension but also retention, since students are more likely to remember and apply concepts they have actively explored and understood.

Thermal Energy Standards Connections and Progressions

The model lesson illustrates numerous Standards in practice. The Standards are seamlessly intertwined so that content (Disciplinary Core Ideas [DCIs]) are learned by using Science and Engineering Practices (SEPS) and Crosscutting Concepts (CCC) (NGSS Lead States, 2013). In addition, the standard's connection and progression illustrate that units can target bundles of Standards. In this lesson, targeted Standards are mostly physical science but also relate to earth and science concepts.

Also, as a cautionary note, if you have a vertical curriculum aligned with considerable consistency, you may worry about using the same activity across grade levels. Please do not fret or become apprehensive. It is common for students to say they have "seen the activity before" yet have little conceptual understanding or can base their ideas on data that serves as evidence for their active meaning-making. When students make these assertions, probe them for reasons for their thinking and whether they can explain concepts they have seen before and support their ideas with evidence to determine the best course of action. Students' scientific rules based on evidence can dictate where to go next and why.

Science Conceptual Understanding (DCIs)

- Energy can be moved from place to place by moving objects or through sound, light, or electric currents (3–5: PS).
- Energy is spontaneously transferred out of hotter regions or objects and into colder ones (MS: PS).
- Temperature is a measure of the average kinetic energy of particles of matter. The relationship between the temperature and the total energy of a system depends on the types, states, and amounts of matter present (MS: PS).
- Variations in density due to variations in temperature and salinity drive a global pattern of interconnected ocean currents (MS: ESS).
- Energy cannot be created or destroyed, but it can be transported from one place to another and transferred between systems (HS: PS).
- The tendency of uncontrolled systems to constantly evolve toward more stable states—that is, toward more uniform energy distribution (objects hotter than their surrounding environment cool down)—reflects the principles described in the second law of thermodynamics (HS: PS).

Science and Engineering Practices (SEPS)

- Analyzing and Interpreting Data
- Developing and Using Models
- Constructing Explanations
- Obtaining, Evaluating, and Communicating Information
- Planning and Carrying Out Investigations (*secondary only*)

Crosscutting concepts (CCCs)

- Patterns
- Cause and Effect
- Scale, Proportion, and Quantity (*secondary only*)

References

American Association for the Advancement of Science (AAAS). (1993). *Benchmarks for science literacy*. Oxford University Press.

Brown, P., & Keeley, P. (2023). *Activating student ideas: Linking formative assessments to instructional sequence in grades 6–8*. NSTA Press.

Driver, R., Guesne, E., & Tiberghien, A. (1985). *Children's ideas in science*. Open University Press.

Driver, R., Squires, A., Rushworth, P., & Wood-Robinson, V. (1994). *Making sense of secondary science: Research into children's ideas*. Routledge.

Erickson, G. L., & Tiberghien, A. (1985). Heat and temperature. In R. Driver, E. Guesne, & A. Tiberghien (Eds.), *Children's ideas in science* (pp. 52–84). Open University Press.

Harris, D. (1981). *Concepts in secondary school physics*. Heinemann Educational Books.

Keeley, P., Eberle, F., & Tugel, J. (2007). *Understanding student ideas in science* (Vol. 2). NSTA Press.

McTighe, J., & Silver, H. (2020). *Teaching for deeper learning: Tools to engage students in meaning making*. ASCD.

NGSS Lead States. (2013). *Next generation science standards: For states, by states*. National Academies Press. www.nextgenscience.org/next-generation-science-standards

Module 2

Rethinking Learning

Purpose: To identify research-based attributes of learners and learning that apply to *Explore-Before-Explain* teaching.

Desired Results

Curriculum designers and teacher leaders will understand that:

★ Adult learners need opportunities just like students to engage in inquiry that builds on their prior knowledge, helps develop more research-oriented mindsets to teaching, and invites them to connect learning to meaningful goals and valuable actions.
  ○ An abundance of scholarship and modern learning theory should be used to support our practice.
  ○ Research can be used to evaluate our notions about what is effective or ineffective at promoting long-lasting understanding.

Module Design Goals: In this module, you will explore contemporary research in cognitive, neuroscience, and developmental psychology that explains the best environments for promoting robust learning experiences. The end product will be a more sophisticated view of learners and learning, the importance of sequence of instruction, and the implications for instructional design from a more pedagogical perspective.

You should work on Module 2 if you have not already considered or identified what matters most in classroom environments in terms of the scholarship on learning. In addition, the research section allows you to establish criteria for adopting a pre-existing curriculum resource.

I start with a model lesson like thermal energy in professional learning to spur educators to think about the best learning environments. Teachers and leaders can use the model lesson to think about teaching and learning from a more pedagogical perspective and reflect on ideas inherent in the idea catcher questions that preceded the model lesson related to salient features in the research:

- ★ *Why is it important to activate students' ideas at the beginning of the lesson?*
- ★ *How does targeting relevant conceptual ideas enhance student motivation?*
- ★ *Why are students' evidence-based claims a key destination for learning?*
- ★ *How does seamless and purposeful assessment enhance learning?*
- ★ *Why are teacher enhancements needed to develop a deeper understanding?*
- ★ *Why is Explore-Before-Explain so powerful for learning and learners?*

Suppose you are working through this section as a professional learning community. In that case, it may be beneficial to divide the sections: Developmental Psychology, Neurosciences, Cognitive Sciences-Learner Centered, Cognitive Science-Knowledge Centered, Cognitive Sciences-Assessment Centered, and Science Education Research. Firstly, it can simplify the task of exploring the research while giving teachers ownership of developing their expertise. Secondly, substantial overlap between the research exists, strengthening the argument about the importance of using scholarship to inform practice.

Teacher teams can create anchor charts that paraphrase key research ideas that address, "What do you notice about the research," "What do you wonder about the research," and "What research data can be connected to aspects of the model lessons?" The anchor charts should be kept for future PLC meetings and can help ground why rethinking learning from a more pedagogical perspective is critical.

Modern Learning

Imagine this: even before the first school bell rings, our youngest minds already burst with knowledge and harbor innate intellectual traits rivaling scientists and statisticians. Yes, it is true! Young children possess a treasure trove of fundamental principles across various science disciplines—physical sciences, life sciences, Earth sciences, and space sciences. They are budding scientists and explorers armed with the essential tools needed to generate knowledge and understanding, all through the magic of scientific practices.

As educators, we have an extraordinary opportunity. We can tap into these intuitive inquiry skills that students bring to school, leveraging their existing knowledge, skills, and understanding as a powerful launching pad for instruction. It is about hands-on experiences with data, where evidence becomes the fuel for creating a solid foundation of knowledge. When students have firsthand learning experiences by doing, they can better construct robust conceptual understanding. Teachers build and sustain students' thinking abilities and empower them to transfer knowledge to new and uncharted territories. Our students do not enter school as empty slates waiting to be filled with knowledge. Instead, they arrive with unique ideas about science and the practices underpinning reliable and valid scientific understanding.

Developmental Psychology

Young children and babies unravel the mysteries of the world around them. In these early years before school, we witness remarkable transformations in a child's ability to comprehend, analyze, and thrive. No matter their age, children are natural explorers, ready to tackle complex challenges. Equally captivating is the way their understanding and learning abilities are shaped by their unique experiences.

The aim is not to delve into exhaustive details about the cognitive variances within these few preschool years but to explore the common threads that weave through different age groups. Research into early learning cognition reveals that children's play is akin to the experimentation of scientists (Gopnik et al., 1999). Young minds constantly probe natural occurrences, question the world's inner workings, and collect valuable information. They engage in scientific thinking, formulating hypotheses tested against evidence, with each piece of evidence refining their theories. Children rely on pattern recognition and cause-and-effect relationships in this process, unlocking deeper insights. This is an iterative hypothesis testing, involving complex calculations with conditional probability. From a statistical perspective, young children learn to gauge the likelihood of events based on others, all at a rapid pace. Their learning is dynamic: new understanding builds upon prior data-rich experiences, allowing them to draw remarkably insightful conclusions from sparse information.

In developmental psychology, we have moved far beyond the notion that children follow rigid stages of intellectual development, gaining specific abilities only with age (Duschl et al., 2007). The literature now celebrates the innate curiosity and problem-solving abilities of students, as well as the pivotal role that active, supportive instruction plays in their learning (Bransford et al., 2000). Consider this: even the tiniest babies, unable to talk or walk, engage in reasoning and critical thinking through their tactile experiences and data analysis to explore their environment. The world is dynamic, and movement offers valuable clues about how objects behave and interact, touching upon fundamental physics concepts. Imagine a baby playfully rolling a ball across the room's floor. In that simple act, there is a world of learning happening. The room, with its carpeted and uncarpeted areas, walls, furniture, and other toys, becomes a playground for discovery. Babies instinctively pay attention to this wealth of information, piecing together their understanding of the environment, often needing more adult guidance. Insights from developmental research reveal that our youngest learners possess the qualities we cherish in all students: a scientific way of thinking characterized by their innate ability to seek patterns and causal relationships in their quest to understand the world.

The social structures, including formal schooling, that influence young children's development of ideas persist as they age. Before school, learning revolves around play, exploration, and informal family interactions. While families remain crucial, teachers (Hattie, 2008) and peers (Rubin et al., 2006) grow in complexity and significance as influencers. Academic expectations rise, and the quality of the school environment, focused on educational standards and benchmarks, impacts student learning (Marzano, 2003; OECD, 2013). These factors shape students' learning abilities and their beliefs about those abilities.

Carol Dweck's (2006) work on "mindset" has profoundly impacted education. Her concepts of "fixed mindset" and "growth mindset" have been woven into teaching practices, curriculum development, and educational policies worldwide. Dweck shows that students often move along a continuum from a fixed to a growth mindset based on their experiences with peers, schools, and teachers. In a fixed mindset, students believe abilities and intelligence are static. In a growth mindset, they believe that effort and perseverance can develop abilities, leading to better learning outcomes. Teachers can encourage this by praising effort rather than inherent ability and framing challenges as opportunities to grow. While social structures play a prominent role in learning, active experiences continue to influence how middle and high school students learn best (Prince, 2004).

If you're wondering whether the impacts of active learning, social interactions, and educational systems extend beyond K-12 to postsecondary education, they do (see Pascarella & Terenzini, 2005; Strayhorn, 2012). Despite the changes one experiences transitioning from K-12 to postsecondary settings, active learning remains the most effective method. In one of the largest meta-analyses of undergraduate STEM education, Freeman et al. (2014) analyzed 225 studies and found that active learning significantly boosts student performance in science, engineering, and mathematics. This study indicated that active learning improved exam performance by almost half a standard deviation—a considerable increase amounting to half a letter grade—and that lecturing raises failure rates by 55%.

Neurosciences

This ever-evolving field explores the intricate relationship between our brains and the process of learning. As we delve into this area of scholarship, we must consider how the incredible brain development over time informs how we design instruction. Not too long ago, the limitations of our technology held back our understanding of the mind's inner workings and how we think and learn. However, the advent of cutting-edge imaging technologies, such as structural and functional MRIs, has ushered in a new era of discovery. Science has moved beyond the once-prevailing notion that most brain development occurs in the first few years of life. Instead, we now enjoy a deeper comprehension of how the brain evolves across our entire lifespan.

This revolutionary progress in the field of neuroscience has profound implications for education. While there is ongoing debate about the extent to which new neurons generate (since most are present at birth), it is clear that the brain reaches nearly 90% of its adult volume by age 6. Within this intricate organ, the connections between neurons, aptly named synapses, shape our knowledge structures (Stiles & Jernigan, 2010). Each neuron, a fundamental building block of the brain, acts as both a receiver and a transmitter of information within the nervous system. Synapses, on the other hand, are the vital bridges or junctions between nerve cells, facilitating the passage of information. In early childhood, essential knowledge structures form in the brain, a process deeply rooted in neurobiology. From a neurobiological standpoint, early learning catalyzes connecting neurons through synapses. It is fascinating that these synapses are not fixed entities; they adapt and change over time in response to experiences and new inputs. Synaptic proliferation is most active in infancy and childhood, slowing down as adolescence unfolds and extending into adulthood (Blakemore, 2010; Blakemore & Choudhury, 2006).

Like tending to a garden, actively used synapses grow more robust, while those that remain inactive or less helpful are pruned away. This pruning process is akin to trimming a rose bush, in that weaker branches are removed to allow the stronger ones to flourish. Synaptic pruning can streamline neural circuits, enhancing knowledge and skills as individuals enter adulthood. However, even as neural pruning may decrease after adolescence, adults can draw upon different neural mechanisms and leverage their wealth of experiences to solve problems and continue learning.

In addition to these insights, recent advances in neuroscience shed light on the chemical processes in the brain that influence learning. This knowledge has profound implications for how we approach instruction and sequencing. For instance, we now understand that all learning commences as sensory information, filtering into distinct brain structures known as the "thinking brain" and the "reactive brain." In the neurosciences, the "thinking" and "reactive" brain refer to different functional aspects of decision-making and behavior. The "thinking brain" is mainly linked to higher-order cognitive process like reasoning, problem-solving and decision-making and allows for deliberate, conscious thought. Instructional practices that engage students' thinking empower them to make predictions, uncover patterns and causal relationships, and view learning as a developmental journey—an inherently metacognitive process. Such practices also foster feelings of pleasure, satisfaction, and motivation, effectively directing information to the "thinking brain." The "reactive brain" is primarily associated with emotional responses, fear, and threat detection. Conversely, instructional approaches that induce stress or anxiety in students, such as the fear of speaking in front of peers or worrying about academic demands, shift the brain into survival mode, channeling input to the "reactive brain" (McTighe & Willis, 2019). These two different functional aspects of the brain are mentioned because how we engage student thinking can "turn on" or "turn off" their learning.

In summary, early learning experiences and the dynamic evolution of the brain shape future learning endeavors, forming fundamental connections within this remarkable organ. Insights from neuroscience reverberate throughout the realm of education as we discover how learning environments can sculpt the physical structure and functional organization of the brain itself.

Cognitive Science

The cognitive sciences help uncover the secrets of how we learn best and how to create optimal learning environments. *How People Learn* (Bransford et al., 2000) and *How People Learn II: Learners, Contexts, and Cultures* (NASEM, 2018) describe integral aspects of top-notch classroom instruction: the learner, knowledge, and assessment.

Learner-Centered: At the heart of this principle lies the idea that knowledge is not something passively absorbed. It is actively constructed through experiences that deeply embed concepts, fostering long-lasting understanding. Fundamental to the idea of learner-centeredness is the idea that all knowledge is constructed through active experience. Active meaning-making takes center stage here. Learners do not just remember facts; they reconstruct and apply conceptual knowledge. They connect the dots within a broader framework, creating a rich understanding. So, it is not just about retrieving specific facts but also about grasping the bigger picture. In addition, learner-centered environments nurture metacognition, empowering students to think about their ideas, monitor their progress, and become more self-reliant learners. This seems like a simple idea, but in fact, it is not. The best active, learner-centered experiences provide experiences that deeply entrench ideas and promote long-lasting understanding. In this regard, long-lasting understanding is highlighted by an individual's ability to reconstruct and apply conceptual knowledge rather than retrieval of specific facts (National Academies of Science, Engineering and Medicine (NASEM), 2018, p. 4). Long-lasting understanding is promoted when learners construct knowledge, connect details within a broader framework for understanding, and relate information with the knowledge they already have.

Finally, the learner-centeredness of an environment is influenced by the upbringing of the individual. How learners grow and learn is related to their cultural, social cognitive, and biological contexts (National Academies of Science, Engineering and Medicine [NASEM], 2018). Learning is influenced in fundamental ways by the student's home environment and the difficulties they encounter may be a mismatch between a student's cultural experiences and the expectations at school. Thus, classroom environments should be learner-centered to highlight activating students' prior knowledge and engaging students in active experiences that promote thinking about developing understanding.

Knowledge-Centered: The knowledge students learn should fit into a broader framework. In knowledge-centered classrooms, we focus on what students should know and how it connects to overarching ideas and equips them with valuable skills. We introduce knowledge components when they are most relevant and beneficial for learners. Educational reforms over the years have set clear standards for what students should learn. These standards align with current research and emphasize the importance of introducing concepts promptly when needed. Knowledge-centered environments guide students in learning that leads to enduring understanding and empowers them to apply their abilities across various contexts.

Assessment-Centered: Effective teaching is only complete with a strong assessment component. Assessment-centered classrooms prioritize high learning standards and provide frequent feedback. Feedback comes in various forms, helping students self-monitor their progress at every stage. It is an ongoing dialogue, a constant navigation toward improved understanding. Formative feedback helps students identify areas of confusion and what to work on next. Summative feedback assesses overall understanding. Assessment-centered classrooms empower students to identify the knowledge and skills they need to reach their full potential.

These principles (learner, knowledge, and assessment) are intertwined, forming a vibrant classroom learning culture. They interact and overlap, creating a harmonious balance for the most effective teaching and learning environments (see Figure 3.1).

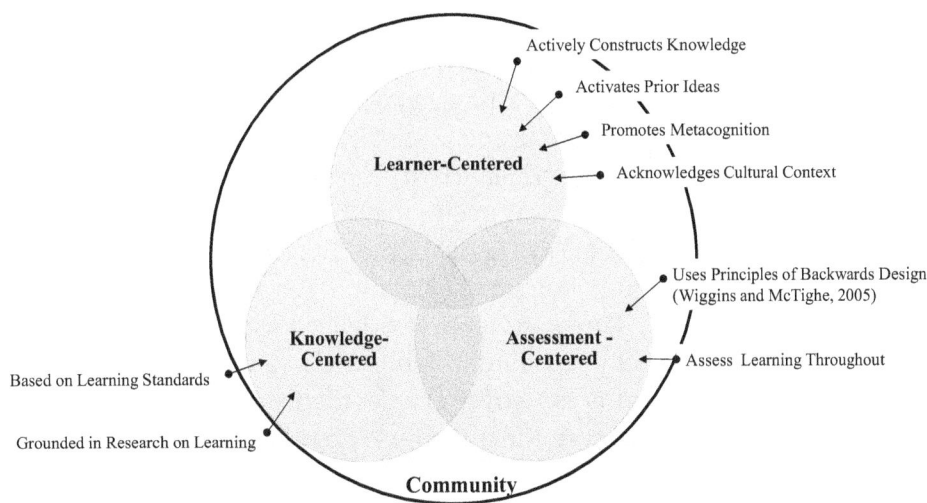

Figure 3.1 Components and their attributes of how students learn best (Bransford et al., 2000; NASEM, 2018)

Science Education Research

Leaders should be well versed in the key features of a coherent, content-rich curriculum that offers a schedule of "what to teach and when" and is inspired by how students learn best. Instructional sequences that address questions inherent in "what to teach and when" have a long history in science education but emphasize that *Explore-Before-Explain* instructional sequences place the experience first and the explanation second. *Explore-Before-Explain* sequences can be traced back to the Science Curriculum Improvement Study (SCIS) curriculum, which is divided into three phases: (1) exploration, (2) invention (concept introduction), and (3) discovery (concept application) (Atkin & Karplus, 1962; Karplus & Their, 1967). Since the learning cycle, the 5E instructional model (see Bybee et al., 2006) provides significant advances directly based on cognitive science research that demonstrates:

> *An alternative to simply progressing through a series of exercises that derive from a scope and sequence chart is to expose students to the major patterns of a subject domain as they arise naturally in problem situations. Activities can be structured so that students are able to explore, explain, extend, and evaluate their progress. Ideas are best introduced when students see a need or a reason for their use—this helps them see relevant uses of the knowledge to make sense of what they are learning.*
>
> (Bransford et al., 2000, p. 127)

Explore-Before-Explain teaching provides a carefully planned sequence of instruction that places students at the center of learning. Designing

Explore-Before-Explain lessons, or implementing them through a purchased curriculum, is less of an argument about using specific approaches like inquiry, project/problem-based learning, discussions, readings, and lectures, and more about there being a right time and place for different approaches according to how students learn best and Standards. Said differently, the approaches implemented are activity- and resource-dependent, but the sequence is critical for promoting learning. A key innovation of *Explore-Before-Explain* is that students' explorations allow them to use data that serve as evidence for their active meaning-making. Thus, students' evidence-based claims—i.e., their explanations—are naturally a Standards-minded way to teach (see NRC 2012). The only way students can arrive at an evidence-based claim about the content (i.e., a disciplinary core idea) is to use scientific practices and crosscutting concepts like pattern recognition and causal relationships.

If we use the scholarship in science education to inform practice, we can expect to see what Michael Fullan (2010) calls "stunningly powerful consequences" for students learning from *Explore-Before-Explain* instruction gleaning insight from research on the 5Es and learning cycle instructional sequences:

- Large-scale meta-analyses show that 5E experiences influence student achievement, conceptual understanding, heightened motivation, and engagement compared to traditional approaches (Bybee et al., 2006; Cakir et al., 2017; Austin et al., 2023).
- For example, Austin et al. (2023) found that the weighted average effect for interventions based on the 5E or related instructional models was +0.82 standard deviations, which is equivalent to a 29 percentile point relative gain for students who learned through the 5E model.
- Research reveals that the 5E does not just nudge the needle; it propels students forward with such a dynamic force that some soar months ahead of their peers in traditional classroom atmospheres (Bybee et al., 2006).
- The 5Es are not just about making exceptional learning for some students but about closing the gap for those left behind (Wilson et al., 2019). The approach is incredibly impactful for English language learners (ELL) (Estrella et al., 2018), students with disabilities (Vavougios et al., 2016), and students from low socioeconomic conditions (Tan et al., 2023).

While the bulleted list may seem exhaustive, it underscores a critical point—designing a curriculum or choosing an established resource must provide the innovations needed to realize the vision of the Standards and promote long-lasting conceptual understanding for students.

Call to Action

Modern standards and research create a golden opportunity to enhance student achievement nationwide. However, it is more complex than adopting modern standards and nudging teachers toward instructional shifts: it requires an enhanced view of the role of scholarship informing practice. All these developments in the study of learning have led to a new understanding of teaching and why instructional sequence matters for learners. As illustrated earlier, new theories about why instructional sequence matters are emerging. Before school even begins, young children already know a surprising amount about their world and have innate intellectual characteristics similar to those of scientists and statisticians. They know basic principles in many different science disciplines (physical sciences, life sciences, and Earth and space science) and the beginnings of valid and reliable ways to generate knowledge and understanding (using scientific practices). If students bring intuitive inquiry skills to learn about their world to school, their teachers can capitalize on what the students know, can do, and understand as a starting point for instruction. In addition, if students prune out weak knowledge structures so the strong ones can flourish, their teachers must focus instruction on activities in which students' firsthand experiences with data that serve as evidence help them construct sound conceptual understanding. Finally, supportive environments nurture, build, and sustain students' thinking abilities so they can transfer knowledge to new situations. Students do not come to school as blank slates ready to be inscribed with expertise. Instead, they have ideas about science and the practices used to generate reliable and valid scientific understanding.

References

Atkin, J. M., & Karplus, R. (1962). Discovery or invention? *Science Teacher*, 29, 45–47.

Austin, M., Polanin, J., Taylor, J., & Steingut, R. (2023, May). *The effects of the 5E instructional model: A systematic review and meta-analysis* [Conference presentation]. American Educational Research Association (AERA) 2023 Virtual Conference.

Blakemore, S.-J. (2010). The developing social brain: Implications for education. *Neuron*, 65(6), 744–747. https://doi.org/10.1016/j.neuron.2010.03.004

Blakemore, S.-J., & Choudhury, S. (2006). Development of the adolescent brain: Implications for executive function and social cognition. *Journal of Child Psychology and Psychiatry*, 47(3–4), 296–312. https://doi.org/10.1111/j.1469-7610.2006.01611.x

Bransford, J., Brown, A., & Cocking, R. (2000). *How people learn: Brain, mind, experience, and school*. National Academies Press.

Bybee, R. W., Taylor, J. A., Gardner, A., Van Scotter, P., Powell, J. C., Westbrook, A., & Landes, N. (2006). *The BSCS 5E instructional model: Origins, effectiveness, and applications*. BSCS. https://media.bscs.org/bscsmw/5es/bscs_5e_full_report.pdf

Cakir, M., Guven, D., & Koksoy, S. (2017). Effects of 5E learning cycle model on students' academic achievement and attitudes towards science. *International Journal of Science Education*, 39(7), 773–793. https://doi.org/10.1080/09500693.2017.1296988

Duschl, R. A., Schweingruber, H. A., & Shouse, A. W. (Eds.). (2007). *Taking science to school: Learning and teaching science in grades K–8*. National Academies Press.

Dweck, C. S. (2006). *Mindset: The new psychology of success*. Random House.

Estrella, J. G., Au, S. M., Jaeggi, S., & Collins, P. (2018). Is inquiry science instruction effective for English language learners? A meta-analytic review. *AERA Open*, 2(10). https://doi.org/10.1177/2332858418761234

Freeman, S., Eddy, S. L., McDonough, M., Smith, M. K., Okoroafor, N., Jordt, H., & Wenderoth, M. P. (2014). Active learning increases student performance in science, engineering, and mathematics. *Proceedings of the National Academy of Sciences*, 111(23), 8410–8415. https://doi.org/10.1073/pnas.1319030111

Fullan, M. (2010). *All systems go: The change imperative for whole system reform*. Corwin Press.

Gopnik, A., Meltzoff, A., & Kuhl, P. K. (1999). *The scientist in the crib: Minds, brains, and how children learn*. William Morrow.

Hattie, J. (2008). *Visible learning: A synthesis of over 800 meta-analyses relating to achievement*. Routledge.

Karplus, R., & Their, H. D. (1967). *A new look at elementary school science*. Rand McNally.

Marzano, R. J. (2003). *What works in schools: Translating research into action*. ASCD.

McTighe, J., & Willis, J. (2019). *Upgrade your teaching: Understanding by design meets neuroscience*. ASCD.

National Academies of Sciences, Engineering, and Medicine (NASEM). (2018). *How people learn II: Learners, contexts, and cultures*. National Academies Press. https://doi.org/10.17226/24783

National Research Council (NRC). (2012). *A framework for K–12 science education: Practices, crosscutting concepts, and core ideas*. National Academies Press.

OECD. (2013). *PISA 2012 results: What makes schools successful? Resources, policies and practices (Volume IV)*. OECD Publishing.

Pascarella, E. T., & Terenzini, P. T. (2005). *How college affects students: A third decade of research*. Jossey-Bass.

Prince, M. (2004). Does active learning work? A review of the research. *Journal of Engineering Education*, 93(3), 223–231. https://doi.org/10.1002/j.2168-9830.2004.tb00809.x

Rubin, K. H., Bukowski, W. M., & Parker, J. G. (2006). Peer interactions, relationships, and groups. In W. Damon & R. M. Lerner (Eds.), *Handbook of child psychology* (6th ed., Vol. 3, pp. 571–645). John Wiley & Sons.

Stiles, J., & Jernigan, T. L. (2010). The basics of brain development. *Neuropsychology Review*, 20(4), 327–348. https://doi.org/10.1007/s11065-010-9148-4

Strayhorn, T. L. (2012). *College students' sense of belonging: A key to educational success for all students*. Routledge.

Tan, C. Y., Gao, L., Hong, X., & Song, Q. (2023). Socioeconomic status and students' self-efficacy. *British Education Research Journal*. https://doi.org/10.1002/berj.3869

Vavougios, D., Tselios, N., & Ravanis, K. (2016). The use of the 5E instructional model in combination with the educational software "Interactive Physics" in the field of Newtonian dynamics. *Journal of Educational Technology & Society*, 19(3), 129–141.

Wilson, C. D., Taylor, J. A., Kowalski, S. M., & Carlson, J. (2019). The relative effects and equity of inquiry-based and commonplace science teaching on students' knowledge, reasoning, and argumentation. *Journal of Research in Science Teaching*, 47(3), 276–301. https://doi.org/10.1002/tea.20313

Module 3

Exploring Sinking and Floating

Purpose: To apply new ideas about learners and learning from the scholarship to content-focused and standards-aligned experiences for ongoing coaching, support, and reflection on the essential elements of *Explore-Before-Explain* learning. Cognitive science research shows the active of actively using information in new and different situations, strengthens understandings (National Academies of Sciences, Engineering, and Medicine (NASEM), 2018).

Desired Results

Curriculum designers and teacher leaders will understand that:

- ★ Effective instructional design can be evaluated using research on how students learn best.
  - ○ Evaluating lesson effectiveness should correspond to the most salient aspects of cognitive science, neuroscience, and developmental psychology research.

Module Design Goals: In this module, you will perform an *Explore-Before-Explain* lesson that takes no more than 50 minutes to connect research to an *Explore-Before-Explain* sequence. The end product will be a more sophisticated view of the learning attributes associated with the importance of the sequence of instruction and the implications for instructional design.

You should work on Module 3 if you are accustomed to traditional teaching methods emphasizing teacher explanations and unfamiliar with the research supporting *Explore-Before-Explain* teaching. In addition, you should work on Module 3 if you teach this content and are vetting the purchase of a pre-established curriculum to ensure it will meet students' needs and modern science education standards.

You might skim or skip this Model 3 and use the thermal energy transfer example to unpack the ideas soon to come.

I invite you to explore another model lesson with two video resources with key research ideas in mind. The model lesson is provided to build your understanding of the research and show how students' curiosity and questions about conceptual science ideas can help develop an understanding of active meaning-making. As you progress through the model lesson, remember the "idea catcher" reflection questions (shown in the next box) and how critical research findings are present in the model lesson. In addition, reflection questions are embedded throughout the model lessons. The embedded reflection questions can be used in groups or individually, highlighting the most salient aspects of instructional design that leverage high levels of active meaning-making.

Reflection Alert! Idea Catcher
- ★ *How are students' ideas activated at the beginning of the lesson?*
- ★ *How does the lesson target relevant conceptual ideas?*
- ★ *How does student exploration lead to students' evidence-based claims?*
- ★ *How are assessments incorporated?*
- ★ *What teacher enhancements are needed to develop a deeper understanding?*
- ★ *How is Explore-Before-Explain being used?*

Materials Needed for This Lesson

- ◆ "Watermelon and Grape" formative assessment probe (included)
- ◆ 10-gallon fish tank
- ◆ 8 gallons of water
- ◆ Watermelon
- ◆ Grape
- ◆ 100-ml graduated cylinder
- ◆ Electronic balance
- ◆ Graduated cylinder

Safety Notes

1. Have direct adult supervision while you are working on this activity.
2. Wear sanitized indirectly vented chemical splash safety goggles meeting ANSI/ISEA Z87.1 D3 standard and non-latex aprons during the activity's setup, hands-on, and takedown segments.
3. Quickly wipe up spilled or splashed water off the floor so it does not become a slip-and-fall hazard.
4. Keep water-filled aquarium away from electrical receptacles to prevent accidental shock.
5. Make sure electrical receptacles are GFI protected.
6. Do not taste or eat food used in this activity.
7. Follow procedures set up by the instructor for disposal of materials.
8. Use caution when working with glassware or plasticware; it can break and cut/puncture skin.
9. Wash your hands with soap and water after completing this activity.

Vignette: Exploring "Sinking and Floating"

The lesson started when students were asked to create a rule for their thinking and guided with questions such as, "What patterns can you observe when comparing objects of different sizes and weights?" To build on these initial ideas, the activities incorporated the *Uncovering Student Ideas* probe "Watermelon and Grape" (Brown & Keeley, 2023; Keeley, 2013, pp. 49–52). The probe asks students whether they think a watermelon and a grape will sink or float. The ideas attracted students; most thought the watermelon would sink and the grape would float. When asked to explain their thinking, students' responses focused on using the object's weight to determine whether it would sink or float. Students' ideas were collected using a sticky chart. A sticky chart asks students to put their predictions on a Post-it note that is then used to create a bar graph of the class ideas. In addition, students were asked to create a rule for their thinking. Many students explained that "the watermelon will sink because it's heavy," and the "grape will float because it's light." When used with secondary students, some students wrote "density." When secondary students were asked how density influences sinking and floating, they resorted to answers only about the weight of the objects, such as, "It is denser because it is heavier." Students who wrote density, as a rule, used only the object's weight to clarify their thinking.

> **Student Ideas Alert!**
> **Here are common student ideas identified in the research.**
> - ★ A study by Biddulph and Osborne (1984) asked students ages 7 to 14 why things float. The typical response was, "Because they are light."
> - ★ Some students use an intuitive rule of "More A–More B" (Stavy & Tirosh, 2000). The reason is that the larger an object, the more likely it is to sink.

> **Reflection Alert! What key research findings support eliciting students' ideas and experiences?**

The probe aimed to elicit student ideas about sinking and floating and see whether they could substitute a prediction with a rule for their thinking. Therefore, engagement time was not about the correctness of students' ideas but about how they logically supported their predictions. The probe revealed a research-identified commonly held intuitive rule, known as "More A–More B," in which students' reason that if there is more of one thing, then there is more of another. For example, if an object has more weight or is larger, students believe it is more likely to sink.

Next, with ideas in mind, it was time for students to explore whether the watermelon and grape would sink or float. We performed the investigation as a classroom demonstration using a 10-gallon fish tank (see video resource here: *www.youtube.com/watch?v=L_Pc6sxB_gI&t=3s*). First, the grape was placed in the tank. To the student's surprise, the grape sank and went straight to the bottom of the tank. Next, we put the watermelon in the tank. Most students were shocked when the watermelon did not sink to the bottom. Thus, the exploration provided students with data that would serve as evidence for scientific claims and help them build a deeper conceptual understanding of why some objects sink while others float in water.

> **Reflection Alert! What key research findings support students constructing evidence-based claims?**

Just for Elementary Students

Next, we set out to do some additional tests. The goal was less about having students identify whether objects would float or sink and geared more toward having them think about objects in terms of whether they are heavy or light. The exploration focused on the idea that weight and size alone do not determine whether objects sink or float. First, students extended the formative assessment probe and compared other fruits, including cranberries and a pumpkin. They compared the relative heaviness of the objects and then predicted whether each would sink or float based on what they had learned from the "Watermelon and Grape" investigation. Students closely compared the grape and cranberry. While nearly the same size and similar in shape, many thought the cranberry would float because it was "lighter." Students were less sure about comparing the pumpkin and watermelon but believed they felt the same and commented that the pumpkin should float like the watermelon. Then, the students watched the demonstration and saw that the pumpkin floated like a watermelon. Thus, the experience added further evidence that weight alone does not determine whether an object sinks or floats.

Further objects were added to the students' investigation, including a small marble, weight, and water balloon. Students immediately thought the marble and weight would sink, basing their explanation on a difference between these materials and the fruits tested. They believed these objects did not have water and thus would sink. Students' conceptions were verified when both the marble and weight sank. Students were less sure about the water balloon. Most thought it would sink because the water balloon felt heavy. Some students thought the water balloon would be suspended in the middle of the tank because, as they explained, "It was water just like the water in the tank."

> **Reflection Alert!** What research findings support using enhancements to develop student understanding further?

The explain phase began with students' evidence-based claims described during class discussions. They had two similar experiences that served as evidence for their claims. Firstly, students consolidated their data into two categories, one on each half of their sheet, with the headings "Heavy for its size" and "Light for its size." Next, they drew pictures of each item tested and listed under one of the headings whether it would sink or float. Students made the following evidence-based claims. Firstly, an object's weight alone does not determine whether it sinks or floats. Students could support this claim with the watermelon and pumpkin versus the grape and cranberry. Secondly, shape alone does not explain whether an object sinks or floats. Students' claims were supported with evidence that even similar-shaped objects like apples, avocados, and tomatoes differ in whether they sink or float. With teacher guidance, students began to explain sinking and floating as related to an object's size and shape. Aiden explained that a "small object that feels heavy will sink like a small rock." While students could do most of the evaluation activity independently, they needed help with the watermelon and pumpkin. All students thought these two objects were heavy and did not have a frame of reference for whether they were heavy for their size. Still, Charlie remarked that even "really big objects can float if they are as big in size as the watermelon."

Thus, students started thinking about heaviness in a relative sense and realized that not all objects perceived as heavy float in water. The culminating activity allowed students to see patterns across the different experiences as evidence that shape and size do not determine whether objects sink or float. Having students summarize all their experiences on one sheet allowed them to understand the similarities between the various floating and sinking experiences.

> **Reflection Alert! What research findings support helping students explain ideas to develop student understanding further?**

To further enhance student ideas, students did related elaborations. The elaboration started with the *Uncovering Student Ideas* probe "Sink or Float?" (Keeley, 2013, pp. 45–48). The probe asks students to consider three different student ideas about whether clay floats or sinks. Students' ideas represented the persistence of beliefs about sinking and floating being related to the shape and size of an object. Most students (91%) agreed with the argument presented by Bonita and thought that "sometimes clay sinks and sometimes it floats," depending on whether the clay is flattened like a pancake. One student thought it was less about the shape and more about whether the clay was soft or hardened. Only a few students (9%) thought it would sink regardless of the shape.

> **Reflection Alert! How are assessments used to develop elaborations?**

Before testing whether different shapes influence sinking or floating, students created the three shapes illustrated in the *Uncovering Student Ideas* probe: round ball, rolled pencil, and flat pancake. They used the same amount each time and weighed the clay. All students found that the clay had the same weight regardless of shape. Since all students received the same small amount of clay, their weight measurements were similar, approximately 3.7 grams.

The class headed outside and used a small baby pool to test the probe. Firstly, all the students tried a clay ball, which sank. Next, they stretched and rolled their clay into a pencil-like shape and placed it in the water. The pencil-shaped clay sank as well. Finally, students flattened their clay into a pancake shape. To their surprise, and contrary to their prediction, this clay also sank. (See teacher video resource at *https://youtu.be/WlqD_1qJovo*.)

Next, students explored whether the properties of other materials influence whether they sink or float. The goal was to have students compare properties like size, weight, and shape to understand what makes objects of certain materials float or sink. For this exploration, all students received different-sized objects made of steel, such as a paper clip, a spoon, and a wrench. They predicted whether the object's weight would affect whether it would sink or float. For example, the students predicted that the spoon and wrench would sink because they were "heavy for their size" but thought the paper clip might float. They tested the objects and found that all three sank.

The teacher then showed them a picture of a large cargo ship with a steel hull and asked them to think about how it could float even though it was made from the same material as the steel objects they had just tested. How was the boat's shape different from the clay shapes and steel objects they had previously tested? Could they create a boat model using clay to show how a material that sinks can float? Students were asked to draw their models of clay boats and then test them using the clay. They weighed the clay so that they used the same amount of clay as in the other tests, changing only the shape. The students discovered that turning up the sides of the flattened clay allowed it to float. The teacher then asked them to sink their floating clay boats and observe whether they floated back up. The students observed that even though they had changed the shape so the boat could float, the same shape could also sink.

Students made overarching scientific claims based on data they used as evidence for active meaning-making. Firstly, they noticed a pattern in the clay weight data. Next, students used multiple individual data points as evidence to claim that (1) weight does not change when we alter an object's shape and (2) weight alone does not determine whether an object sinks or floats. Finally, they connected what they had observed with their clay boats to how a steel tanker ship can float.

Next, students needed to learn that shape alone does not determine whether an object sinks or floats. The students were puzzled at how the same shape that floated when the sides were turned up like a boat could also be made to sink. The teacher pressed them to consider the difference between the floating clay boat shape and the same shape that sank. The students came up with the idea that when the boat was floating, it had air in it, but when the water got inside the boat, that made it heavier, causing it to sink. The other sinking objects were not shaped so that they could hold air. Therefore, the shape made a difference when it allowed the object to be filled with a lighter material, such as air. There was a cause-and-effect relationship between the shape and the material: if the shape could hold air, the material could float.

> **Reflection Alert! How do students' evidence-based claims further develop their conceptual understanding?**

Finally, of the four shapes of clay tested, students could explain why three sank and one floated. With teacher guidance, students began conceptualizing whether an object sinks or floats as influenced by the combination of observable physical properties such as shape, size, weight, and material.

Students revisited the "Watermelon and Grape" and "Sink or Float" formative assessment probes to explain that size and shape determine whether an object sinks or floats. Students supported their scientific explanation with evidence that small objects that are heavy for their size, like a grape, will sink and large objects that are light for their size, like a watermelon, can float. (Please encourage students to think using relative comparisons, because they need a frame of reference for considering whether an object is heavy or light for its size.) While students revised their claims, they were prompted to think about sinking and floating based on observations such as size, shape, and materials that make up the object rather than just one factor alone.

> **Reflection Alert! How do students' evaluation and developing understanding lead to more profound meaning-making?**

Just for Secondary Students

The explain phase began with students finding the mass and volume of the grape using water displacement. In three groups, students were supplied with a 100-ml graduated cylinder, a grape, water, and an electronic balance. Students easily collected data on the grape's mass. They were more torn on how to use the materials to find the volume of the grape but had some prior experience using graduated cylinders to measure volume for other labs. Students realized that when the grape was placed into the graduated cylinder, water "moved up," and the value was greater than before. Students made a simple calculation of the water level before and after the grape was placed in the graduated cylinder to measure the volume of the grape. We turned

Table 4.1 Each group's grape's mass and volume

Group	Mass (g)	Volume (ml)	Density (g/ml)
1	5	4	1.25
2	5.5	4.5	1.22
3	5.1	5	1.02
4	5	4.9	1.02
5	4.9	5	0.98
6	5.4	5	1.08

"little" into "big" data and created a table on the front board that included each group's grape's mass and volume (see Table 4.1). In addition, we discussed as a group what density meant on a conceptual level as the relationship between an object's mass (m) and volume (v) and that density (d) could be calculated using $d = \frac{m}{v}$. Each student calculated density (see Table 4.1).

Students looked for patterns in the class data to claim the density of a grape. They noticed that in three out of five cases, the calculated density of the grape was greater than 1. They also questioned the accuracy of each other's measurements. While it was pretty straightforward to measure the mass of the grape, many groups commented that it was difficult to measure the volume of the grape using water displacement. As a class, we learned that some groups rounded their measures up or down depending on the water line (meniscus) without a consensus across the class. Students thought that the average density value might be the best analysis since there was variation in the group's measurement techniques (average density was equal to 1.10 g/ml). Thus, students could use data as evidence that a grape's density is greater than 1.

Next, we discussed how we could use a similar process to find the density of a watermelon. While we could find the mass (mass equaled 2,870 g), we did not have graduated cylinders or beakers big enough to find the volume of the watermelon. We decided to use the fish tank because the water level would rise due to the watermelon. Because the fish tank was a rectangle and regular solid, students could use mathematical calculations for volume based on prior math and science experiences (length × height × width). We also noticed that because the watermelon floated, if we gently pushed on it to submerge it completely, then the water level would rise even higher. As a result, we measured the water level before and after the watermelon was completely submerged. Students used the values for the change in height, length, and width to find the column of the watermelon using water displacement (volume equaled 3,156.25 cm²). Based on the mass and volume of the watermelon, students calculated the watermelon's density to be 0.909 g/cm³. As a result, students could make the evidence-based claim that a watermelon's density is less than 1.00.

The last investigation challenged students to find the density of water to help tie together their classroom experiences and better understand the scientific explanations behind the initial formative assessment probe. Students used prior experiences with the graduated cylinder and electronic balances to first take the mass of the graduated cylinder. Next, they filled the graduated cylinder with water and took the mass. When students subtracted the mass of the graduated cylinder, they found that the mass of the water always equaled the volume. Hence, when students calculated density, they found a value of 1.00 g/ml. Finding the density of water allowed students to have a meaningful comparison of when objects sink or float. Objects having a density less than that of water float, while objects having a density grater than that of water sink.

> **Reflection Alert! Consider how the enhancements lead to deeper student meaning-making. What research supports using enhancements?**

The teacher's explanation focused on physical properties and occurred as a class discussion. We discussed the shortfalls in describing materials based on size, shape, or weight. For example, although we say an object is big or small, that description does not provide us with information about the object. Instead, comparing the object's mass, volume, and density provides valuable insights into the material. It opens the conversations that would soon come about other physical properties such as melting, boiling, and freezing, and the points at which they occur.

Students revisited the "Watermelon and Grape" formative assessment probe to explain that the relationship between mass and volume (density) determines whether an object sinks or floats. Students supported their scientific explanation with evidence that small objects that are heavy for their size, like a grape, will sink and large objects that are light for their size, like a watermelon, can float. In addition, students used quantitative density measurements to justify their thinking and revise their rules. Even students who initially wrote "density" as their rule were able to elaborate and use data as evidence to explain how knowing an object's density can help determine whether it will sink or float.

> **Reflection Alert!** How does assessment promote students' reflection on developing understanding?

Unpacking the Exploring Sinking and Floating Model Lesson

These lesson elements targeted in the idea catcher and their sequence highlight emerging research in developmental psychology, neurosciences, and cognitive sciences. In this section, I unpack the idea catcher questions.

How Are Students' Ideas Activated at the Beginning of the Lesson?

In the model lesson, students' ideas are engaged at the onset of the unit using the "Watermelon and Grape" formative assessment probe. Nearly all students have experience with objects sinking or floating. Thus, the probe taps into students' background ideas and situates learning in an exciting and thought-provoking situation. Three expansive bodies of scholarship—developmental psychology, neurosciences, and cognitive sciences—support that all students bring a wealth of experience to science classrooms. Teachers can use students' past experiences as assets and can use students' ideas to activate learning. This approach aligns with the constructivist theory of learning, which posits that students build new knowledge on the foundation of their existing understanding. By activating prior knowledge, students are better prepared to integrate new information, making the learning experience more meaningful and relevant.

How Does the Lesson Target Relevant Conceptual Ideas?

The relevancy of the assessment probe resides in its ability to promote questioning about whether different objects sink or float, which is directly based on their everyday experiences. The targeting of relevant conceptual ideas makes learning purposeful. Regarding research in developmental psychology, neurosciences, and cognitive sciences, students are more motivated to learn about and remember ideas that are everyday experiences in their lives. For example, students have experienced watching small twigs and large logs float down the river. Students can connect these experiences to their understanding of why watermelons float while grapes sink. These lived experiences can lead to deeper discussions about density and density relationships (which involve both mass and volume), concepts that are critical in various scientific disciplines. This relevance not only piques students' curiosity but also enhances their engagement and retention of the material.

How Does Student Exploration Lead to Students' Evidence-Based Claims?

The "Watermelon and Grape" demonstrations provide data that supports their active meaning-making. The demonstration is vital and serves as the focal point of the lesson. Students are problem solvers and question-answerers who innately use pattern recognition and cause-and-effect relationships to understand their world (supported by developmental psychology, neurosciences, and cognitive sciences). Students use patterns and cause-and-effect relationships to make evidence-based claims about science content, namely, density relationships to sinking and floating, highlighted in standards (e.g., pure substances have physical properties that can be used to identify them). For instance, after observing the watermelon and grape in water, students might hypothesize that the size or mass of an object determines whether it sinks or floats. Through further exploration and discussion, they refine their understanding to focus on density, ultimately constructing more accurate and scientifically valid explanations.

How Are Assessments Incorporated?

Assessments are used seamlessly throughout the lesson and serve many purposes. Firstly, assessments like the "Watermelon and Grape" formative assessment probe create relevancy and activate student thinking at the lesson's onset. Teacher-student discourse focuses on asking students to think about patterns and cause-and-effect relationships and developing thinking versus introducing content and validating students' ideas. From a research perspective, students making predictions about science creates a flood of neurobiological activity. Students want to verify that their predictions are accurate. This intrinsic motivation drives deeper engagement and persistence in learning tasks. The conversations that occur and revisiting initial ideas are also a formative assessment and promote students thinking about their developing understandings. These discussions allow students to articulate their thinking, receive feedback, and refine their ideas, fostering a deeper understanding. Finally, students reflecting on their developing understanding helps them become more self-reliant learners and identify how using data as evidence for budding ideas can promote more accurate conceptual understanding. This metacognitive aspect of learning is crucial because it empowers students to take control of their learning process and apply their knowledge to new situations.

What Teacher Enhancements Are Needed to Develop a Deeper Understanding?

Teacher enhancements allow students to use scientifically accurate terminology in light of their firsthand experiences (for secondary students, the density equation). Fundamental research in neurosciences and cognitive science supports introducing ideas in light of students' firsthand experiences. Just-in-time learning is a rich experience because students internalize new terms and concepts related to their developing conceptual frameworks that they have constructed firsthand. For example, after students observe and discuss the density relationships to sinking and floating objects, the teacher might introduce the concept of density, explaining how the mass-to-volume ratio determines density relationships (which involve both mass and volume). This timely introduction of terminology helps students connect their concrete experiences with abstract scientific concepts, solidifying their understanding. Furthermore, enhancement activities might include additional experiments or simulations that allow students to apply and test their newly acquired knowledge, deepening their conceptual grasp.

How Is *Explore-Before-Explain* Being Used?

This two-day lesson on density relationships to sinking and floating illustrates the importance of sequencing science instruction according to how students learn best. During each phase of the lesson, there is considerable consistency between *Explore-Before-Explain* and research, as well as confronting limits in understanding, setting the stage for where to go next. The lesson starts by engaging students' ideas related to common instances. This engagement phase not only captures students' interest but also provides a diagnostic tool for the teacher to assess prior knowledge and misconceptions. Next, students explore specific ideas so they have data that serves as evidence for active meaning-making. This exploration phase is critical for allowing students to engage in hands-on, minds-on activities that foster inquiry and discovery. Then, students construct evidence-based claims. This process of claim construction encourages students to think critically and justify their ideas with evidence, aligning with scientific practices. Following students' evidence-based claims, the teacher provides enhancement-type activities to sophisticate understanding and ensures targeted Standards are addressed. This phase might include explicit instruction, guided practice, or collaborative discussions that deepen students' understanding and help them integrate new knowledge. Throughout the lesson, students consider their developing understanding and how they use practices and crosscutting concepts to understand science better. This reflective process helps students make connections between different concepts and see the broader applicability of their learning.

The *Explore-Before-Explain* approach is rooted in the belief that students learn best when they are actively engaged in the learning process, constructing their own understanding through inquiry and investigation. This method contrasts with traditional didactic approaches where information is presented up front, often disconnected from students' experiences and prior knowledge.

If students are allowed to explore first, they develop a deeper, more meaningful understanding of scientific concepts, which is then solidified and extended through targeted explanations and enhancements. This approach not only aligns with how students naturally learn but also prepares them to think and act like scientists, fostering a lifelong curiosity and passion for science.

By incorporating these strategies and approaches, teachers can create a dynamic and engaging learning environment that promotes deep and lasting understanding of scientific concepts. This method empowers students to become active participants in their learning journey, developing critical thinking, problem-solving, and evidence-based reasoning skills that are essential for success in science and beyond.

Exploring "Sinking and Floating" Standards Connections and Progressions

The model lesson illustrates numerous Standards in practice. The Standards are seamlessly intertwined so that content (Disciplinary Core ideas [DCIs]) are learned by using Science and Engineering Practices (SEPS) and Crosscutting Concepts (CCC) (NGSS Lead States 2013).

Conceptual Science Understandings (DCIs)

- Matter can be described and classified by its observable properties. Different properties are suited to different purposes. (K-2: PS)
- Measurements of a variety of properties can be used to identify materials. (3–5: PS)
- Each pure substance has characteristic physical and chemical properties that can be used to identify it. (6–8: PS)
- Variations in density due to variations in temperature and salinity drive a global pattern of interconnected ocean currents. (6–8: ESS)

Science and Engineering Practices (SEPs)

- Carrying Out Investigations
- Analyzing and Interpreting Data
- Constructing Scientific Explanations

Crosscutting Concepts (CCCs)

- Patterns
- Cause and Effect
- Structure and Function

References

Biddulph, F., & Osborne, R. (1984). *Pupil's ideas about floating and sinking* [Paper presentation]. Australian Science Education Research Association Conference, Melbourne, Australia.

Brown, P., & Keeley, P. (2023). *Activating student ideas: Linking formative assessments to instructional sequence in grades 6–8*. NSTA Press.

Keeley, P. (2013). *Uncovering student ideas in primary science: 25 new formative assessment probes for grades K–2*. NSTA Press.

National Academies of Sciences, Engineering, and Medicine. (2018). *How people learn II: Learners, contexts, and cultures*. The National Academies Press. https://doi.org/10.17226/24783

NGSS Lead States. (2013). *Next generation science standards: For states, by states*. National Academies Press. www.nextgenscience.org/next-generation-science-standards.

Stavy, R., & Tirosh, D. (2000). *How students (mis)understand science and mathematics: Intuitive rules irrespective of their correctness*. Teachers College Press.

Module 4

How Can We Make Science Learning Remarkable?

Purpose: To illustrate how innovation is sometimes less about creating something all new and more about rethinking our current structures.

Desired Results

Curriculum designers and teacher leaders will understand that:

- ★ We can scale up our practices by identifying what works best in the classroom.
- ★ Homing in on our hands-on, minds-on experiences allows us to identify other areas for improvement in classroom lessons.

Module Design Goals: To engage teachers in thinking about the possibilities that exist within their current practices. The end product will be a view that innovation is within our abilities and can have lasting and dramatic impacts on students.

You should work on Module 4 if you want to motivate your teachers to see that impactful changes do not always require significantly new and different practices.

I hope, at this point, you are finding a renewed sense of purpose and realizing that instructional design matters. Using novel examples outside education as motivators and thought-provokers can spur additional insights into what is possible in our classes. Sometimes, we learn from the success of others to spark the change we may

> *need to take students to new levels or desired by our teams. This short section on making learning remarkable honors our expertise and how change can promote a positive, creative process to the art of teaching.*

Whether you use an iterative approach (altering something that currently exists) or an inventive approach (creating something new), the goal is to create a better instructio nal sequence for students by starting with exploratory experiences before explaining science content. By rethinking our approach to instructional design, we can foster a deeper understanding and appreciation of science among students. This guidebook will explore various instructional design considerations and ideas to enhance your teaching methods, drawing inspiration from innovative practices in other industries.

One effective analogy for understanding the importance of innovation in instructional design is to consider the changes that have occurred in the paint industry. Let's start by thinking about the traditional one-gallon paint can. For many years, this type of container was the standard in the industry. However, when you think about the traditional paint can, several issues come to mind. How comfortable is the thin metal handle? Not very comfortable, especially when carrying one or more paint cans over a long distance. The thin metal handle can dig into your hand, making the task quite uncomfortable.

Next, consider how easy it is to remove the lid using a "key" or screwdriver. The process can be quite cumbersome and often results in bent lids or, worse, a struggle that can lead to spills. This brings us to another point: what sort of mess is made when pouring paint from the can? Traditional paint cans are notorious for drips and spills, which can be frustrating and time-consuming to clean up. These issues and others prompted the need for change in the paint industry.

In 2002, an American paint company set out to make their product genuinely remarkable and introduced an all-plastic, twist-and-pour container with a twist-off lid, side handle, and pouring spout to reduce drips and spills. This new packaging design addressed many of the pain points associated with the traditional paint can. Simply changing the packaging had a tremendous impact on the user experience. Dutch Boy Paint's sales tripled in three months! By switching from the traditional container to a more user-friendly design, Dutch Boy was able to change consumers' conceptions about paint (see *www.partisanadvertising.co.nz/blog/three-steps-to-increasing-sales*).

This example from the paint industry illustrates an important point about innovation: sometimes, it is less about changing what works (in this case, the paint itself) and more about changing the packaging. The paint remained the same, but the way it was presented and used by consumers was transformed.

This change in packaging made the product more accessible, easier to use, and ultimately more appealing to consumers.

Similarly, in science education, innovation can often be less about changing the hands-on experiences we use and more about highlighting the most salient features of how students learn best in a purposeful instructional sequence. Reflecting on innovation and iteration can help you plan engaging lessons that tap into your experience and expertise. Your experiences are critical in rethinking instructional design to promote active meaning-making among students.

To further illustrate this point, let's consider a few examples of how you can apply innovative thinking to your instructional design. One approach is to start with a hands-on, exploratory activity that captures students' interest and curiosity. For instance, the thermal energy lesson started with students' ideas. Next, learning progressed to provide students with firsthand experience with data that served as evidence for their sensemaking. This simple yet engaging activity can serve as a hook to draw students into the lesson and spark their interest in learning more about the underlying science.

Once students have had the opportunity to explore and observe, you can then introduce more formal instruction to explain the science content behind the activity. By starting with an exploratory experience, you provide students with a concrete example that they can relate to the more abstract concepts being taught. This approach helps to make the learning more meaningful and memorable.

Another strategy is to incorporate analogies and real-world examples into your lessons. Just as we used the example of the paint industry to illustrate the importance of innovation, you can use similar analogies to help students make connections between the science content and their everyday lives. For example, when teaching about the concept of density, you might use the analogy of oil and water separation in a salad dressing to help students understand why some substances float while others sink.

Additionally, consider the use of technology to enhance your instructional design. Tools such as interactive simulations and digital models can provide students with opportunities to explore scientific concepts in new and engaging ways. These technologies can also offer ways to differentiate instruction, allowing you to tailor lessons to meet the diverse needs of your students.

As you reflect on these strategies and examples, it's important to remember that your experiences and insights as a teacher are invaluable. This guidebook is meant to honor your abilities and experiences, recognizing that you bring a wealth of knowledge and expertise to your instructional design. You want to maintain your good practices and have strategies to shore up less effective ones when creating a more robust learning environment for students.

In conclusion, whether you are using an iterative approach to refine your existing practices or an inventive approach to create new ones, the goal is to design instructional sequences that start with exploratory experiences and build toward a deeper understanding of science content. By drawing inspiration from innovative practices in other industries, reflecting on your experiences, and incorporating engaging and meaningful activities into your lessons, you can create a more effective and enjoyable learning experience for your students.

Module 5

Planning Engaging and Effective Science Lessons

Purpose: To prioritize and focus our lessons on the essential elements of active meaning-making using an *Explore-Before-Explain* sequence.

Desired Results

Curriculum designers and teacher leaders will understand that:

★ Effective instructional design prioritizes learning around active student experience.
  The innovations required by the standards create meaningful lessons that:
  ○ Activate student ideas
  ○ Allow students to construct knowledge using the three dimensions of the frameworks
  ○ Enhance student understanding so they can transfer ideas to new and different situations
  ○ Allow students to become more self-directed learners through reflecting on what works best for their developing understanding

Module Design Goals: In this module, you will learn how to priority-plan *Explore-Before-Explain* lessons that seamlessly combine the three dimensions of the *Frameworks*. The end product will be a curriculum- and lesson-planning strategy to maximize student learning.

> **You should work on Module 5 if** you have a homegrown curriculum (i.e., you created your own curriculum) or have the ability to adapt adopted curriculum resources to better meet student needs. You should also work on this model if you are vetting a pre-designed curriculum to ensure it aligns with the instructional design practices associated with how students learn best.

The opportunity for immense and immediate progress becomes clear when we take an unblinking look at the possibilities in our classrooms. Please take to heart that this book's model lessons and research chapter point to instructional design practices designed to promote active meaning-making. Considering the model lessons and research chapters may shed light on areas of your teaching and curriculum that are ripe for improvement—reorienting our focus through simple shifts in instructional sequence and clarity about instructional priorities. Belinger (1992) suggested using "big variables" when planning for curricular change. Big variables, research-based elements that make classroom experiences powerful for students, are the core of active meaning-making through *Explore-Before-Explain*.

Brace yourself. In the following section, I offer four essential features of active meaning-making ("framework") and suggestions for curriculum planning that apply *Explore-Before-Explain* to act on this idea. These ideas have previously undergone peer review and include my previous collaboration with several individuals (see Brown & Bybee, 2024; Brown et al., 2023a; Brown et al., 2023b; Brown & Bybee, 2023a, 2023b). Active meaning-making is the intentional process of designing instruction so students construct and assign meaning to experiences and information. It emphasizes the active role of prior knowledge and experiences in learning and creating an understanding of common instances and conceptual ideas. Active meaning-making goes beyond passively receiving information and encourages students to analyze and interpret information and consider new understandings in light of past experiences. We can prioritize planning the four essential features of active meaning-making with the activities we use in a powerful *Explore-Before-Explain* instructional sequence. Promoting active meaning-making through *Explore-Before-Explain* empowers individuals to construct their understanding of science concepts.

Although there are many ways to start planning, beginning with specific outcomes in mind brings greater leverage to the overall lesson, unit, and curriculum design. When planning, consider the principles of how students learn best, along with their subcomponents. The lesson elements that follow and their sequence underscore a fundamental point about achieving deep and lasting learning—students need to actively "make meaning" to come to understand abstract science concepts (McTighe & Silver, 2020).

Step 1: Plan "Backward" (Constructing Claims)

Instead of beginning with traditional lessons and favorite activities, using Understanding by Design (UbD) procedures helps educational leaders identify the lesson and unit learning outcomes in terms of what would count as acceptable evidence and how students would have experiences to allow them to make more accurate scientific claims. UbD emphasizes that units should focus on transfer goals that specify what students should do with their learning in the long run (Wiggins & McTighe, 2005).

Planning backward involves identifying the desired evidence-based claims that students should be able to make at the end of the unit. This approach ensures that teaching is focused on developing students' conceptual understanding and their ability to transfer learning through purposeful science investigations. By starting with the end goals in mind, teachers can create a coherent and content-rich curriculum that aligns with the goals of the Frameworks for K-12 Science Education and the Next Generation Science Standards (NGSS) (National Academies of Sciences, Engineering, and Medicine, 2022; NGSS Lead States, 2013; NRC, 2012).

The Importance of Evidence-Based Claims

There is simply too much science content to cover it all within the limited hours of the school day. Research shows that without a coherent focus, the curriculum can become a series of disjointed lessons that amount to "hundreds of hours of wasted class time each year" (Kane & Steiner, 2019). Using the Standards can help finely focus the content on what is most essential for understanding. Therefore, starting with the end goals for student learning in mind allows all other planning considerations to emerge from this focused approach.

Focusing on evidence-based claims is a powerful way to teach in a standards-minded manner. Students can only arrive at an evidence-based claim through the unique combination of Disciplinary Core Ideas (DCIs), Science and Engineering Practices (SEPs), and Crosscutting Concepts (CCCs). By nesting the appropriate Science and Engineering Practices and Crosscutting Concepts within the lesson, teachers can ensure that students develop a deep understanding of the underlying concepts using practices and logical thinking. This is crucial if we want students to use science to understand and explain their world. As Michael Fullan (2005) states, "Terms travel easy . . .

but the meaning of the underlying concepts does not" (p. 67). Empowering students to do the hard intellectual work and use critical and logical thinking to construct conceptual knowledge is essential for meaningful learning.

Focusing on evidence-based claims in science education not only aligns with but also can help operationalize the goals of performance expectations, ensuring that students engage in authentic science practices that deepen their understanding and prepare them for real-world problem-solving. While not all performance expectations (PEs) explicitly require students to make evidence-based claims as the central focus, this skill remains a foundational aspect of science education. For instance, when evaluating models, testing designs, or describing phenomena, students often need to justify their conclusions or choices based on observations, making the practice of evidence-based reasoning valuable. Said a bit differently, and to illustrate with a specific example, while many PEs focus on students developing or using models, the importance of evidence-based claims is critical for the assertions students construct from model-based thinking, indirectly involved in justifying their model choices, and communicating their findings. Thus, pinpointing the evidence-based claims students can make and applying them to more specific PEs can help teachers bridge the *Frameworks* to practice.

Interdisciplinary Benefits

Pinpointing evidence-based claims as evidence of learning also offers considerable interdisciplinary benefits. For example, students' evidence-based claims can serve as assessment evidence for informative/explanatory writing, as advocated by the Common Core State Standards (CCSS) in English Language Arts (ELA). Additionally, students often need to use mathematical practices such as measurement and data analysis, which are endorsed by the CCSS in Mathematics (National Governors Association Center for Best Practices & Council of Chief State School Officers, 2010). Connecting science to other areas of learning not only develops a deeper understanding in authentic contexts but also promotes transfer learning, which is particularly valuable when a teacher's time may be limited (NASEM, 2022).

Practical Implementation in the Classroom

Implementing backward planning in the classroom involves several practical steps. Firstly, teachers need to clearly define the learning objectives and the evidence that will demonstrate students have achieved these objectives. This requires careful consideration of the DCIs, SEPs, and CCCs that are most relevant to the unit of study. In other words, while standards are not the curriculum, they are key to instructional design to ensure students are gaining scientific literacy.

Designing Learning Experiences

Once the learning objectives and assessments are in place, teachers can design learning experiences that will help students achieve the desired outcomes. These experiences should be engaging and relevant, drawing on students' prior knowledge and experiences. Inquiry-based learning, where students investigate scientific questions and develop their own evidence-based claims, is an effective approach. Hands-on experiments, group projects, and real-world problem-solving activities can all help students develop a deeper understanding of scientific concepts. Encouraging students to think about how their learning applies to the real world can help them see the relevance and importance of science in their everyday lives.

> Big Idea #1: Make students' evidence-based claims the focus of your active meaning-making lesson design. All other essential elements of active meaning-making and the seamless integration of the three dimensions of the Standards should emerge from the goal of having students construct an evidence-based claim.

> **Strategies for Big Idea #1: Plan Backward, Targeting the Evidence-Based Claims Students Can Construct**
>
> ★ Use an existing problem-solving situation, demonstration, or simplified lab.
>   - Use what you have and know works.
>   - You can still teach procedures and should teach safety.
>   - A key destination in planning is students' evidence-based claims.
> ★ Talk with students about data evidence claims science principles.
>   - Help students formulate clear lines of arguments.
>   - Remember, when students construct evidence-based claims, they naturally experience important interdisciplinary learning called for in ELA and mathematics.

In conclusion, planning backward using the UbD framework is a powerful approach to designing science instruction. By focusing on the desired learning outcomes and the evidence that will demonstrate student understanding, teachers can create a coherent and meaningful curriculum that promotes deep learning and transfer. This approach aligns with the goals of the NGSS and other educational standards and offers significant interdisciplinary benefits. While it may require a shift in thinking and a significant amount of effort to implement, the benefits for student learning make it a worthwhile endeavor.

Step 2: Activating Students' Ideas: Creating a Need-to-Know Situation

Once we have targeted the evidence-based claims students can make about science, the next critical step is to decide how to engage their ideas effectively. Beginning new lessons with students' ideas and experiences is a powerful strategy that fosters a "need-to-know" situation. This approach not only stimulates curiosity but also activates student learning by connecting new content to what they already know. When students see a direct link between their prior knowledge and the new concepts being introduced, they are more likely to engage deeply and meaningfully with the material.

Phenomena-Based Teaching and the Importance of Relevance in Engagement Activities

Targeting conceptual ideas in our engagement activities ensures that the content is relevant to students' lives. By focusing on occurrences that students encounter in their everyday experiences, we create learning situations that resonate with them on a personal level. Phenomena-based teaching does precisely this because it situates learning around real-world events or occurrences. This relevance sparks curiosity and encourages students to ask questions about their lived experiences. For example, if the lesson is about forces, starting with an activity that involves pushing or pulling objects that students interact with daily will make the lesson more engaging and meaningful. The chosen topic should invoke curiosity and prompt students to explore and question based on their own experiences, thus enhancing their engagement and motivation to learn.

Productive Discourse: Increasing Student Engagement

"Notice" and "wonder" talk and questioning routines offer practical frameworks for fostering productive discourse in the classroom. These strategies help engage students in conceptual discussions and promote equitable learning experiences for all students. For instance, when students are asked, "What do you notice?" they are invited to share their observations without fear of judgment or assessment. This approach encourages all students, including those who may be hesitant to speak up, to contribute their insights based on their experiences. It creates an inclusive environment in which diverse perspectives are valued and considered.

The "wonder" question, such as "What do you wonder about this?" is particularly effective because it focuses students' attention and creates a genuine need to know. McTighe and Willis (2019) highlight that such questions are among the highest-yield instructional strategies because they stimulate curiosity and drive inquiry. By using probing questions like "notice" and "wonder," teachers can reveal students' thinking processes and experiences, which are crucial for gathering evidence and guiding further exploration. Once wonderment questions are identified, students can make predictions based on their prior knowledge and experiences, which can then be tested and explored through hands-on activities.

Discrepant Events: Stimulating Curiosity and Inquiry

Discrepant events serve as another powerful tool for engaging students' ideas. These are hands-on, minds-on experiences that present occurrences contrary to students' natural trains of thought and expectations. Discrepant events do not necessarily need to be extraordinary or spectacular; rather, they should challenge students' preconceived notions and provoke their curiosity. The effectiveness of discrepant events lies in their ability to capture students' natural inquisitiveness and problem-solving abilities. By presenting students with situations that contradict their prior experiences or assumptions, teachers can stimulate their curiosity and motivate them to investigate further. This approach makes learning accessible to all students by tapping into their inherent curiosities and encouraging them to question and explore.

Probing Questions: Activating Thinking and Meaning-Making

Asking students specific probing questions about common scientific occurrences is another effective way to activate their thinking. When students are prompted to reflect on a scientific event and provide reasons for their thinking based on their experiences, they engage in active meaning-making. For instance, if the lesson focuses on the concept of density, asking students why certain objects float while others sink can prompt them to apply their understanding and think critically about the underlying principles.

Using tested probes, such as those developed by Keeley and colleagues, can be beneficial in engaging students' ideas. These probes are designed to elicit students' thinking and reveal their conceptual understanding. Teachers can either use these existing probes or create their own tailored to their specific lesson goals. The key is to design questions that encourage students to draw on their experiences and reasoning, thereby bridging the gap between their prior knowledge and new scientific concepts.

Bridging Learning Experiences

Finally, using students' ideas and experiences at the beginning of lessons helps bridge learning experiences, making it easier for students to make sense of new ideas. By starting with familiar concepts and gradually introducing more complex content, teachers can create a seamless transition that supports students' understanding. This approach ensures that new information is not presented in isolation but is connected to what students already know, enhancing their ability to grasp and retain new concepts.

In summary, engaging students' ideas through relevance, productive discourse, discrepant events, and probing questions creates a dynamic and interactive learning environment. These strategies help activate students' curiosity, stimulate inquiry, and support active meaning-making. By leveraging students' prior experiences and interests, teachers can design lessons that are not only educational but also engaging and meaningful.

> **Big Idea #2:** Students' background experiences and ideas are powerful assets that help contextualize and motivate learning. Using student's ideas bridges old and new understandings to develop deeper conceptual understanding.

> **Strategies for Big Idea #2: Encourage Students' Background Experiences and Ideas to Activate Learning**
>
> ★ Probe student thinking ("What do you notice, what do you wonder?").
> ★ Have students make a prediction and explain their ideas before doing a demonstration or laboratory.
> ★ Use an Uncovering Student Idea (USI) probe (see Page Keeley).
> ★ *Create sticky bar charts:* tally students' incoming ideas so the class can see similarities and differences. Keep students' ideas anonymous. (Keeley, 2008)
> ★ Ask for reasons for thinking ("rules").
> ★ Do not give the answers to the pre-assessment.

Step 3: Enhance Understanding (Connecting Claims to Scientific Principles)

Once students have formulated evidence-based claims and engaged in initial explorations, the next crucial step is to enhance their understanding by connecting these claims to fundamental scientific principles. This step is vital for deepening students' conceptual knowledge and ensuring that their learning is not just superficial but robust and transferable. Enhancing understanding involves several layers of instructional strategy, each designed to address gaps in knowledge and to consolidate and extend learning.

Understanding Enhancement Activities

Enhancement activities serve to refine and deepen students' initial claims and explorations by confronting limitations in their current understanding and guiding them toward more sophisticated scientific concepts. These activities are not merely supplementary but are integral to helping students make connections between their explorations and broader scientific principles. For instance, if students initially explored the concept of buoyancy through floating and sinking objects, an enhancement activity might involve a more detailed investigation into the principles of density and displacement. This not only solidifies their understanding but also helps them apply these concepts to new and varied contexts.

The Role of Scientific Vocabulary and Terminology

A significant aspect of enhancing understanding is the introduction and reinforcement of essential scientific vocabulary and terms. As students progress in their learning, they benefit from being equipped with precise terminology to articulate their ideas effectively and to engage with scientific content at a higher level. For example, in a unit on forces, introducing terms such as "force," "friction," and "gravity" becomes crucial. This vocabulary should be introduced in context, relating directly to students' firsthand experiences and explorations. For instance, after an activity exploring how different surfaces affect the motion of objects, students could engage in a discussion about "friction" and "force," linking these terms to their observations.

Utilizing NGSS Frameworks for Vocabulary and Concepts

The Next Generation Science Standards (NGSS) provide a comprehensive framework for identifying essential vocabulary and concepts. The NGSS outlines DCIs, SEPs, and CCCs, which can guide teachers in selecting the most pertinent terms and concepts for their lessons. For example, if the focus is on the concept of energy transfer, teachers can draw from NGSS to incorporate terms such as "kinetic energy," "potential energy," and "energy transformation." This alignment ensures that the vocabulary and concepts taught are relevant to the standards and are crucial for students' scientific literacy.

Incorporating Readings, Discussions, and Lectures

Enhancing understanding also involves integrating various instructional strategies such as readings, discussions, and lectures. These methods can provide students with deeper insights into the scientific principles underlying their explorations. For example, after engaging in a hands-on activity about thermal energy, students might read a passage about the laws of thermodynamics or participate in a discussion about energy conservation. Such activities help students to connect their empirical findings with theoretical concepts, facilitating a more profound understanding of the subject matter.

Further Explorations as Elaborations

Elaborations in the form of further explorations are another key component of enhancing understanding. These activities allow students to apply their knowledge to new situations, thus promoting transfer learning. For instance, after exploring buoyancy with simple objects, students could be challenged to design and test their own boats to see how well they float under different conditions. This kind of exploration not only reinforces the initial concepts but also encourages students to use their understanding in innovative ways.

Promoting Transfer Learning

Transfer learning is a critical goal of enhancement activities. It involves applying learned concepts and skills to new and varied contexts. For instance, students who have studied the concept of force in a physics unit might later apply this understanding to analyze the forces involved in engineering design projects. Effective enhancement activities should therefore be designed to encourage students to make connections between their learning and real-world scenarios, thereby demonstrating the relevance and applicability of scientific principles.

> **Big Idea #3:** There is way too much science content to cover it all. The Standards can strategically help you know what to introduce in light of students' experiences and where to go next in the storyline for student learning.

> **Strategies for Big Idea #3: Enhance Understanding Through Discussions, Reading, Lectures, and Simulations**
>
> ★ Focus readings/lectures/discussion on "need-to-know" information that promotes enduring understanding.
>   ○ Reading strongly connects to ELA standards for technical/informational experiences (National Governors Association Center for Best Practices and Council of Chief State School Officers (NGAC and CCSSO), 2010).

- ★ Connect need-to-know information to Disciplinary Core Ideas (see NGSS Lead States, 2013).
- ★ If needed, use elaboration-type activities to sophisticate understanding and allow students to test the utility of budding ideas.

Step 4: Promoting Reflection on Learning

Promoting reflection on learning is a critical step in the educational process that allows students to think deeply about what they have learned, assess their progress, and plan for future growth. This process of reflection, often referred to as metacognition, involves students considering their own learning processes and strategies, evaluating their understanding, and setting goals for further improvement. Engaging in metacognitive activities significantly affects learning outcomes and contributes to the development of higher-order thinking skills (Bransford et al., 2000).

The Importance of Metacognition

Metacognition encompasses a range of activities that help learners become more aware of their cognitive processes. This includes reflecting on what they have learned, how they have learned it, and how they can apply it in different contexts. By engaging in metacognitive practices, students develop a deeper understanding of their learning processes and become more adept at self-regulating their learning strategies. Research indicates that students who actively engage in metacognitive activities—such as self-assessment, goal-setting, and reflection—are more likely to achieve academic success and perform better than their peers according to specific standards who do not engage in these practices (Wang et al., 1990).

Effective strategies for promoting reflection in the classroom involve creating opportunities for students to think critically about their learning experiences. Engaging in reflection allows students to critically evaluate their progress and develop a clearer understanding of their strengths and areas for improvement. This self-awareness is crucial for setting realistic learning goals and for making informed decisions about how to approach future learning tasks. Additionally, reflection promotes a growth mindset, encouraging students to view challenges as opportunities for growth rather than as obstacles.

Studies demonstrate that students who consistently engage in reflection and metacognitive activities tend to perform better academically. For example, John Hattie's landmark review, which analyzed over 800 meta-analyses of educational research, identified self-reported grades—now termed "student visible learning"—as the second most influential factor impacting student achievement. This emphasizes the profound effect that reflecting on one's own understanding can have on academic performance. Here are several practical approaches:

Structured Reflection Activities: Incorporate activities such as journaling, self-assessment checklists, and reflective essays into your lessons. For example, after completing a science project, students could write a reflection on what they learned about the scientific method, how their understanding evolved, and what they would do differently in future investigations.

Regular Feedback and Goal Setting: Provide students with regular, constructive feedback on their work and encourage them to set specific, achievable goals for their future learning. This could involve setting goals related to mastering specific concepts or improving their application of scientific practices. For instance, after receiving feedback on a lab report, students might set a goal to enhance their ability to analyze data more critically.

Peer Reflection and Collaboration: Encourage students to engage in peer review and collaborative reflection activities. This could involve students sharing their reflections with a partner or small group, providing feedback to each other, and discussing their learning processes. Peer reflection helps students gain new perspectives on their learning and fosters a collaborative learning environment.

Metacognitive Prompts and Questions: Use specific prompts and questions to guide students in reflecting on their learning. Questions such as "What strategies helped you understand this concept better?" or "How did you overcome challenges during this investigation?" can help students focus their reflection on their learning processes and strategies.

In summary, promoting reflection on learning is a vital component of effective teaching. By providing students with opportunities to engage in metacognitive practices, such as self-assessment, goal-setting, and structured reflection activities, educators can enhance students' ability to understand their own learning processes, set meaningful goals, and achieve greater academic success. These practices not only improve students' current performance but also equip them with the skills necessary for lifelong learning and personal growth.

> **Big Idea #4:** Developing more self-sufficient, independent learners is critical for students to be equipped to solve new and different problems in school and life. Take every chance possible to show students the successes in their abilities as learners.

> **Strategies for Big Idea #4: Promote Student Thinking About Their Developing Understanding**
>
> ★ **Accurately self-assess:** *How well did I perform? What was most challenging? What grade do I deserve?*
> ★ **Reflect on their learning:** *What was most interesting or surprising about this (topic or event)? What strategies worked well for me during this learning experience? I used to think that _____, but now I understand that _____.*
> ★ **Set future learning goals:** *What will I try next time to improve?*

Step 5: *Explore-Before-Explain*

Implementing an *Explore-Before-Explain* approach is a powerful strategy for fostering deep, meaningful learning. This method aligns with the principle that students should first engage in exploration and investigation before receiving direct instruction. By prioritizing exploration, educators allow students to actively construct knowledge based on their own observations and experiences. This approach enhances engagement, deepens understanding, and helps students internalize scientific concepts more effectively.

Activating Students' Ideas

The first step in the *Explore-Before-Explain* sequence is to engage students' ideas through activities that activate their prior knowledge and experiences. This initial engagement is crucial because it helps students connect new information to what they already know. For example, if students are learning about the principles of buoyancy, starting with a hands-on activity in which they predict and test whether various objects will sink or float allows them to draw on their everyday experiences with floating and sinking. By eliciting students' prior ideas, educators create a context for learning that is relevant and meaningful. This strategy not only activates student thinking but also piques their curiosity and sets the stage for deeper exploration.

Exploration and Evidence-Based Claims

Following the engagement phase, students should be given ample opportunities to explore scientific concepts through hands-on activities and experiments. During this phase, students gather data and evidence to construct scientific claims. For instance, after the initial buoyancy activity, students could conduct a series of experiments to measure how different factors (such as the shape or material of an object) affect its buoyancy. By engaging in these explorations, students learn to make evidence-based claims and develop their understanding of scientific principles. This exploration aligns with the Understanding by Design (UbD) framework, which emphasizes the importance of developing conceptual understanding through active investigation. The exploration phase encourages students to think critically, analyze data, and apply their findings to draw conclusions.

Enhancement and Introduction of New Concepts

Once students have engaged in exploration and constructed preliminary claims, the next step involves introducing new concepts and terminology that may not be easily assessed through exploration alone. Enhancements are designed to address gaps in understanding and to provide a deeper, more comprehensive view of the scientific concepts being studied. For example, after students have explored buoyancy, the teacher might introduce terms such as density and displacement to further explain why certain objects float or sink. Enhancements can include direct instruction, readings, discussions, or multimedia resources that help students refine their understanding and integrate new concepts into their existing knowledge framework. These enhancements should be carefully aligned with the exploration activities to ensure coherence and relevance.

Reflection

The final step in the *Explore-Before-Explain* approach is to provide students with opportunities to reflect on their learning and consolidate their understanding. Reflection allows students to think about what they have learned, how their ideas have changed, and what strategies were effective in their exploration. For instance, after conducting experiments on buoyancy, students might write a reflection on how their understanding of buoyancy has evolved, what they found surprising, and how they can apply their new knowledge to other situations. Reflection activities can also include class discussions, self-assessment, and goal-setting for future learning. By reflecting on their learning process, students develop metacognitive skills and gain a clearer understanding of their own learning strategies and progress.

Arranging Elements Into an *Explore-Before-Explain* Sequence

To effectively implement the *Explore-Before-Explain* approach, educators need to carefully sequence these elements—engagement, exploration, enhancement, and reflection—into a coherent instructional plan. This sequence ensures that students have the opportunity to actively engage with scientific concepts before receiving direct instruction. The critical point is that by arranging the essential active meaning-making features in this order, educators create a learning environment that promotes deeper understanding and long-lasting conceptual knowledge. This approach not only helps students build a solid foundation of scientific principles but also encourages them to take ownership of their learning and develop critical thinking skills that are essential for success in science and beyond.

In summary, the *Explore-Before-Explain* approach emphasizes the importance of engaging students' ideas, providing opportunities for exploration, introducing new concepts through targeted enhancements, and promoting reflection on learning. By following this sequence, educators can facilitate meaningful learning experiences that help students develop a deeper understanding of scientific concepts and foster their ability to apply their knowledge in various contexts.

> **Big Idea #6:** *Explore-Before-Explain* offers a purposeful instructional sequence that offers all students the same access and experiences to build on prior knowledge and to learn by doing science.

Conclusions

In closing, it is important to recognize that successful planning in education may not always adhere to a strictly linear progression. Traditional views of lesson planning often envision a straightforward, step-by-step approach, but effective teaching requires a more dynamic and flexible framework. Rather than following a rigid, linear path, educators can benefit from a more iterative and reflective planning process. This means that while we may begin our instruction by focusing on students' initial ideas and prior knowledge, the planning process itself should be adaptive, allowing for adjustments based on ongoing assessments and student feedback.

To elaborate, while initiating instruction with students' ideas is crucial, the real benefit of our planning comes from identifying and focusing on the evidence-based claims that students can make from their experiences. This approach entails recognizing the specific, tangible outcomes we want students to achieve through their learning experiences. By clearly defining these evidence-based claims, educators can create a more targeted and effective instructional plan that aligns with the goals of the curriculum and standards. This method not only enhances the coherence of the instruction but also ensures that students are engaged in meaningful learning activities that directly contribute to their understanding of scientific concepts.

Incorporating this approach into our practices has the potential to drive significant improvements in student learning outcomes. By zeroing in on high-leverage instructional practices—those that have the greatest impact on student learning—we can move beyond incremental changes. High-leverage practices involve strategies and techniques that are proven to have a substantial effect on students' ability to grasp complex concepts and apply their knowledge in various contexts. Examples of high-leverage practices include formative assessments, targeted feedback, and differentiated instruction. These practices help to address individual learning needs, provide timely support, and ensure that all students have the opportunity to succeed.

When educators effectively implement these high-leverage practices, the results can be transformative. As Michael Fullan (2005) aptly describes, such practices can lead to what he terms "stunningly powerful consequences" for students. This refers to the profound and often remarkable improvements in student achievement and engagement that can result from thoughtful and strategic instructional planning. By focusing on evidence-based claims and utilizing high-leverage practices, educators can create a learning environment that not only meets the immediate needs of students but also fosters long-term academic growth and success.

In summary, while planning for instruction is a critical component of teaching, it should not be confined to a linear framework. Embracing a more flexible and reflective approach allows educators to better address students' needs and enhance their learning experiences. By focusing on evidence-based claims and leveraging high-impact instructional practices, we can achieve substantial improvements in student outcomes and create a more effective and engaging learning environment. Ultimately, this approach supports the development of students' scientific understanding and prepares them for continued success in their educational journeys.

References

Berliner, D. C. (1992). Telling the stories of educational psychology. *Educational Psychologist, 27*, 143–161. doi:10.1207/s15326985ep2702_2

Bransford, J. D., Brown, A. L., & Cocking, R. R. (Eds.). (2000). *How people learn: Brain, mind, experience, and school* (Expanded ed.). National Academy Press.

Brown, P., & Bybee, R. (2023a). Promoting sensemaking: An impactful instructional sequence for teaching elementary students whether objects are heavy or light for their science. *Science and Children, 60*(4), 30–33.

Brown, P., & Bybee, R. (2023b). Promoting sensemaking through an impactful instructional sequence. *The Science Teacher, 90*(6), 22–27.

Brown, P., & Bybee, R. (2024). Promoting sensemaking through an impactful instructional sequence. *Science Scope, 90*(6), 22–27.

Brown, P., McTighe, J., & Bybee, R. (2023a). Promoting learning for all through explore-before-explain. *The Science Teacher, 90*(7), 24–27.

Brown, P., McTighe, J., & Bybee, R. (2023b). Leadership matters: Activating student learning through explore-before-explain. *Science and Children, 60*(6), 8–10.

Fullan, M. (2005). *Motion leadership: The skinny on becoming change savvy*. Corwin.

Kane, T. J., & Steiner, D. M. (2019, April 10). Don't close the book on curricular reform. *Education Week, 38*(28), 26–27.

Keeley, P. (2008). *Science formative assessment: 75 practical strategies for linking assessment, instruction, and learning*. Corwin Press.

McTighe, J., & Silver, H. (2020). *Teaching for deeper learning: Tools to engage students in meaning making*. ASCD.

McTighe, J., & Willis, J. (2019). *Upgrade your teaching: Understanding by design meets neuroscience*. ASCD.

National Academies of Sciences, Engineering, and Medicine. (2022). *Science and engineering in preschool through elementary grades: The brilliance of children and the strengths of educators*. The National Academies Press. https://doi.org/10.17226/26215

National Governors Association Center for Best Practices and Council of Chief State School Officers (NGAC & CCSSO). (2010). *Common core state standards*. NGAC and CCSSO.

National Research Council (NRC). (2012). *A framework for K-12 science education: Practices, crosscutting concepts, and core ideas*. The National Academies Press. https://doi.org/10.17226/13165

NGSS Lead States. (2013). *Next generation science standards: For states, by states*. The National Academies Press. https://doi.org/10.17226/18290

Wang, M. C., Haertel, G. D., & Walberg, H. J. (1990). What influences learning? A content analysis of review literature. *Journal of Educational Research, 84*(1), 30–43. https://doi.org/10.1080/00220671.1990.10885988

Wiggins, G., & McTighe, J. (2005). *Understanding by design* (Expanded 2nd ed.). ASCD.

Module 6

Exploring Fall Time

Purpose: To apply the essential elements of active meaning-making and *Explore-Before-Explain* teaching to content-focused and standards-aligned experiences for ongoing coaching, support, and reflection on the essential elements of *Explore-Before-Explain* learning. Cognitive science research shows the power of actively using information in new and different situations strengthens understanding (National Academies of Sciences, Engineering, and Medicine [NASEM], 2022).

Desired Results

Curriculum designers and teacher leaders will understand that:

- ★ Effective instructional includes essential elements:
  - ○ Activating student ideas and experiences
  - ○ Allowing students to make evidence-based claims
  - ○ Enhancing understanding through explanations and further elaborations
  - ○ Reflecting on how ideas change and develop

Module Design Goals: In this module, you will see how the essential elements of active meaning-making and arranged in an *Explore-Before-Explain* sequence show up in classroom practice. The end product will be a more refined understanding of how the essential elements of active meaning-making and *Explore-Before-Explain* teaching appear in different lessons.

> **You should work on Module 6 if** you need further examples of how the essential elements of active meaning-making and *Explore-Before-Explain* teaching play out in practice.
>
> **You might skim or skip this module if** you can pinpoint the essential elements of active meaning-making and *Explore-Before-Explain* teaching in the previous model lessons.

I invite you to explore another model lesson with essential features of active meaning-making in mind. The model lesson is provided to build your understanding of the research and four essential elements of active meaning-making and show how students' curiosity and questions about occurrences in their lives can help develop students' conceptual understanding. In addition, this model lesson takes an inquiry approach to classroom learning. As you progress through the model lesson, remember the idea catcher questions (shown in the next box) and how essential active meaning-making elements are in the model lesson. In addition, reflection questions are embedded throughout the model lessons. The embedded reflection questions can be used in groups or individually and highlight the most salient aspects of instructional design that leverage high levels of active meaning-making.

> ***Reflection Alert! Idea Catcher***
>
> ★ *How are students' ideas activated at the beginning of the lesson?*
> ★ *How does the lesson target relevant conceptual understanding?*
> ★ *How does student exploration lead to students' evidence-based claims?*
> ★ *How are assessments incorporated?*
> ★ *What teacher enhancements are needed to develop a deeper understanding?*
> ★ *How is Explore-Before-Explain being used?*

Materials Needed for This Lesson

- Maple tree seeds or a picture of maple tree seeds
- Paper
- Helicopter template
- Paper clips

Safety Notes

1. Have direct adult supervision while you are working on this activity.
2. Wear sanitized safety glasses with side shields or safety goggles meeting ANSI/ISEA Z87.1 D3 standard when dealing with solids (e.g., meter stick, projectiles, and glassware)
3. The teacher should practice and demonstrate all procedures before having students try them.
4. Wash hands with soap and water after completing this activity.
5. Clear away all fragile items from the activity zone if a moving helicopter becomes a projectile.
6. Quickly pick up any items used in this activity off the floor so they do not become a slip-and-fall or trip-and-fall hazard.

Vignette: Exploring "Fall Time"

The lesson began by showing students maple tree seeds, called Samaras, prompting students to share what they noticed. Students quickly identify a variety of characteristics: some maple tree seeds have longer blades or wings than others, some are misshapen, some are "drier," and some seeds are bigger while others are smaller. In addition, students notice color differences among the seeds. Students' observations naturally lead to wonderment questions. Students begin to wonder about the maple tree seeds' shape, why they have blades and a seed (capsule), and what the maple tree seeds do. To deepen students' curiosity, a maple tree seed is released so students can see it swirl and twist through the air.

> **[Reflection alert! How is a relevant conceptual occurrence being used to activate student thinking?]**

Students use a graphic organizer called the inquiry planning wheel to organize students' ideas about their observations and wonderment questions. The inquiry planning wheel supports students in developing scientific experiments instead of giving students specific directions. Much like a Ferris wheel, the graphic organizer consists of three main parts: the central hub, spokes, and passenger cars. In the lesson, the central hub asks students to think about a specific aspect of some conceptual science topic. The goal is to get students to think about how something works or acts in natural instances. Then, students list factors that could influence what they have written in the "central hub" in the "passenger cars" on the graphic organizer. So, for instance, if students select "fall time," their planning wheel might look as follows (see Figure 7.1).

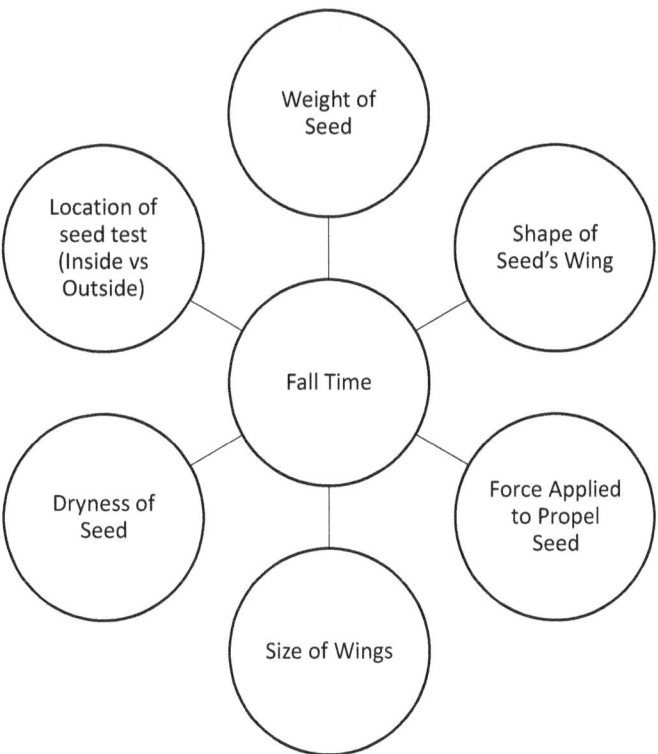

Figure 7.1 Inquiry planning wheel

What the inquiry planning wheel so nicely orchestrates for students is the relationship between different variables in an investigation. Once students complete the planning wheel, the class selects a factor they want to test in the exploration. Guiding questions ask students whether they can change more than one factor simultaneously. These probing questions guide the inquiry-based investigation. For instance, students discuss whether we could change another listed factor simultaneously as we change the size. They quickly realize that the remaining ideas should stay the same throughout the investigation once we decide what to change. According to the graphic organizer, the chosen spoke becomes the independent variable, the center of the planning wheel is the dependent variable, and all other spokes are constants in an investigation. Whether teachers introduce these terms at this grade level or address them on a conceptual level is up to them. Using the planning wheel, students combine the independent and dependent variables to form research questions for classroom investigation. For example, after some debate, students reach a consensus to explore whether the size of the maple tree seed influences fall time. The planning wheel sets the stage as students refine investigations and carry out other related activities.

With ideas in mind, students predict that the size of the maple tree seed may influence flight time, providing rules for their thinking. Many students propose that the larger maple tree seeds would fly faster (hit the ground first) than the smaller ones. These students claim that bigger objects would fall more quickly. Other students note that air resistance influences the fall time for the maple tree seeds. These students use air resistance as the logic to explain that smaller maple tree seeds would hit the ground before larger ones.

Student Ideas Alert!

The following are common student ideas identified in the research.

★ Students do not always identify a force to account for falling objects. They think objects "just fall naturally" or the person letting go of the object has caused it to fall (Driver et al., 1994).

★ Studies by Osborne (1984) found that students think heavier objects fall more quickly.

★ Students, including university students, tend to think that heavier objects fall to Earth more quickly because they have greater acceleration due to gravity (Driver et al., 1994).

Reflection Alert! How are students' ideas and experiences elicited and engaged?

As students begin their exploration, they quickly notice that maple tree seeds have multiple characteristics beyond just size that we had identified in the planning wheel. For example, some big seeds have damaged wings, and some tiny seeds appear drier than others. In addition, the sizes varied on a continuum. Also, students could not control for all of the variability between the seeds. Said differently, they had trouble having enough seeds with one unique characteristic to isolate just one variable. Using a paper-designed helicopter, students use a model to explain and predict the influence of size on fall time (see example template here: *www.jpl.nasa.gov/edu/learn/project/make-a-paper-mars-helicopter/*). The paper helicopter allows students to isolate one testable variable and control for others to explain how size influences fall time. In addition, the model's utility was that it should allow them to predict the approximate fall time for other helicopter sizes outside our test ranges (in other words, students could interpolate and extrapolate data as evidence for active meaning-making).

The last activity of the day asks students to consider the pushes and pulls acting on a helicopter that might influence fall time (note that for grades 3 and beyond, the class uses the term "forces" to describe pushes and pulls). Nearly all students identify that the helicopter was pulled to the ground. Some students also realized the air pushed against the helicopter (e.g., the force of air resistance, "drag") would act in the opposite direction when the helicopter fell. Students are less sure about the relationship between the magnitudes of these pushes and pulls (e.g., forces) and how they may influence different-sized helicopters. Their understanding of how to represent pushes and pulls acting on an object is incomplete, but those explanations were soon to come.

On the second day, students collected data on their helicopters. I offered some support when students had questions and had them work in groups of three (recorder, dropper, and timer). Many questions emerged as students began to conduct their initial tests. As students work, many questions emerge and also prompt further questioning. For example:

Student Question:	"Do we have to drop the helicopters from the same height?"
Teacher Answer:	"How would changing the height you dropped each helicopter influence your results?"
Student Question:	"Do we just drop each helicopter one time?"
Teacher:	"How many times do you think you need to drop each helicopter to get accurate results?"

> **Reflection Alert! How are students prompted to think about ways to learn by doing science?**

The investigation centers on students learning by "doing science" in a logical way and not focused on giving precise directions and affirming ideas. Students experience quite a bit of learning through trial and error about how to experiment. Although many feel uncomfortable without having a standard procedure to follow, they quickly realize their role is to collect data and start to think logically about trends, patterns, and relationships. As students performed their explorations and observed patterns in their data, they became more confident that their approach was reliable and that they could conduct their investigations (see Tables 7.1 and 7.2 for representative student data for grades K-2 qualitative data and 6 and beyond for quantitative data).

Many students commented that the investigation was more straightforward than they had initially anticipated. Through experience, they recognize the importance of controlling as many factors as possible to ensure accurate results after seeing variations in their data due to error. Students also notice that conducting one trial for each helicopter does not give them accurate results; however, ten or more trials for each helicopter mass are also unnecessary. At the end of the exploration, students focus on data collection and developing their abilities as scientists. They use logical thinking skills to make sense of their data when looking for patterns. Once students start discussing their data, the lesson transitions into what the data might mean.

Table 7.1 Students' data in grades K-2

Paper Helicopter Size	Helicopter Fall Time (seconds)		
	Trial 1	Trial 2	Trial 3
Small	Fastest	Fastest	Faster
Medium	Slower	Slower	Slower
Large	Slowest	Slowest	Slower

> **Reflection Alert! How can young children construct evidence-based claims?**

Table 7.2 Students' data in grades 6 and beyond

Paper Helicopter Size	Helicopter Fall Time (seconds)			
	Trial 1	Trial 2	Trial 3	Mean
Small	2.9	2.7	2.7	2.8
Medium	4.7	4.5	4.5	4.6
Large	8.4	8.2	7.9	8.2

> **[Reflection alert! How do students construct evidence-based claims?]**

Students were ready to start making evidence-based claims, allowing the lesson to further delve directly into the content being explored. First, students construct an evidence-based claim that the size of the helicopter influences the time it takes to fall. They also compare their data with their initial prediction as a way for them to think about their knowledge and develop understanding. Next, students go beyond their evidence-based claim to describe the scientific principles involved in the investigation. They examine the pushes and pulls acting on the object, including a downward pull and a push due to air (see video resource here: *https://youtu.be/Aw3EfS38-zw*). Students need technical instruction on representing a study system using push and pull (force) arrows. Their push and pull (force) diagrams were a tool for teaching model-based reasoning because they explained and helped students make predictions about a system of study. For example, the length of an arrow can be used to indicate the magnitude of the force, while the way the arrow points shows the direction. When students look at their initial ideas about the pushes and pulls (forces) involved with helicopters, they realize that helicopter size influences the surface area interacting with air, impacting the forces of air resistance and gravity. For instance, air pushing up on the helicopter (air resistance, drag) acts in the opposite direction and is not equal in magnitude. Students represent the pushes and pulls (forces) acting on the bigger helicopters with longer arrows pointing in the opposite direction as the air (air resistance, drag). Their diagrams show that bigger blades mean more surface area in contact with air and greater pulls (forces) when in contact with the air. Students also become more specific in their push and pull (force) diagrams regarding cause and effect relationships. Thus, the model helps them explain how unbalanced pushes and pulls (forces) cause different motion and fall times for helicopters (NGSS Lead States, 2013).

> **Reflection Alert! What enhancements are needed to achieve Standards-based understanding?**

At the end of the activity, students revisited their initial ideas and predictions about how the size of a helicopter influences fall time. Students reviewed their predictions and constructed scientifically accurate explanations based on data and used push and pull (force) diagrams to explain the underlying scientific principles.

> **Reflection Alert! How are students using assessments to reflect on their developing understanding?**

Just for Students in Grades 3–5 and Beyond

A further enhancement using the investigation planning wheel helps students see other factors that can drive investigations to learn more about paper helicopters. Testing other factors builds students' knowledge to promote deeper understanding. As a class, students decide that helicopter weight (or mass) is an excellent factor to test and will further develop their understanding of fall time (See Table 7.3 for representative student data.)

Just for Secondary Students

The final enhancement activity considers force diagrams in new and different situations. Students begin by predicting whether a feather or a hammer will fall to the ground faster. All students confidently suggest that a hammer falls faster, basing these assertions on their experiences with the paperclip-helicopter exploration. With ideas in mind, students watch a video segment of NASA scientist David Scott dropping a hammer and a feather on the moon (*https://ninepbs.pbslearningmedia.org/resource/phy03.sci.ess.eiu.galmoon/galileo-on-the-moon/*). Students comment that they had been "tricked" but are still surprised that the hammer and feather both fell and hit the ground at the

Table 7.3 Students' data in grades 3–5 and beyond

Number of Paper Clips	Helicopter Fall Time			
	Trial 1	Trial 2	Trial 3	Mean*
1	12.5	11.4	11.4	11.7
2	12.5	10.6	10.9	11.3
3	10.4	10.6	10.0	10.3
4	10.1	10.4	8.9	9.8
5	7.8	7.8	8.0	7.9
6	7.1	7.1	7.0	7.1
7	6.4	6.8	6.5	6.5
8	4.7	4.5	4.5	4.6

*Just grades 6 and beyond

> **Reflection Alert!** How do further explorations help students transfer understanding and generate new meanings in new situations?

same time. They revise their force diagrams to reconsider dropping a hammer and feather on the moon. Students notice that the push forces from air resistance are removed from the diagram; however, the pull forces remain intact. Thus, more massive objects have greater pull forces (which they label gravitational forces) than less massive ones. The idea of the relationship between gravitational forces and mass is pivotal in helping to explain why objects hit the ground at the same time. More gravitational forces are needed to speed up a more massive object than a less massive one, which aligns with a fundamental understanding of Newton's second law.

> **Reflection Alert! How do enhancements achieve Standards-based understanding?**

As a final elaboration and assessment of student learning, students are challenged to design a helicopter with a different size and mass than the one tested as a class—one that would fall in precisely 2.0 seconds given a specific flight distance. The evaluation is an engineering task because students have to design a solution to a problem and generate iterative testing so they can redesign and perfect their model. Students draw their prototypes, including their labeled force diagram (summative assessment), before they conduct tests. In addition, students create force diagrams for their chosen mass. As an exit ticket, students are asked three related questions: (1) What was challenging about the exploration? (2) What grade do they believe they deserved? (3) How can they justify the grade they believe they deserve from their experiences in the lesson?

> **Reflection Alert! How does further exploration allow students to generate new meanings and reflect on their developing ideas?**

References

Driver, R., Squires, A., Rushworth, P., & Wood-Robinson, V. (1994). *Making sense of secondary science: Research into children's ideas*. RoutledgeFalmer.

National Academies of Sciences, Engineering, and Medicine. (2022). *Science and engineering in preschool through elementary grades: The brilliance of children and the strengths of educators*. The National Academies Press. https://doi.org/10.17226/26215

NGSS Lead States. (2013). Next generation science standards: For states, by states. The National Academies Press. https://doi.org/10.17226/18290

Osborne, R. (1984). Children's dynamics. *The Physics Teacher*, 22(8), 504–508.

Unpacking the Helicopter Fly Time Model Lesson

These lesson elements are targeted in the idea catcher, and their sequence highlights the four essential features of active meaning-making. The following are representative answers for the idea catcher questions.

How Are Students' Ideas Activated at the Beginning of the Lesson?

In the model lesson, students' ideas are engaged at the onset of the unit using the inquiry planning wheel. Nearly all students have experience with objects falling. Thus, the inquiry planning wheel taps into students' background ideas and situates learning in an exciting and thought-provoking situation. Teachers can use students' past experiences as assets and can use students' ideas to activate learning. This initial activation is crucial because it aligns with the constructivist approach to education, where learners build new knowledge upon the foundation of their existing ideas and experiences. By starting with something familiar, students are more likely to feel confident and engaged, setting a positive tone for the rest of the lesson.

How Does the Lesson Target Relevant Conceptual Ideas?

The relevancy of the assessment probe resides in its ability to promote questioning about curiosities about maple tree seeds, which is directly based on their everyday experiences. Targeting occurrences in students' lives makes learning purposeful. When students can see the connection between what they are learning in the classroom and their real-world experiences, they are more likely to be motivated and retain the information. In this case, the lesson leverages the natural curiosity students have about why and how things fall, using it as a gateway to introduce more complex scientific concepts. This approach aligns with educational research that emphasizes the importance of contextualizing learning to make it meaningful and relevant to students' lives.

How Does Student Exploration Lead to Students' Evidence-Based Claims?

The helicopter explorations provide data that supports students' active meaning-making. The explorations are crucial and serve as the focal point of the lesson. Students use patterns and cause-and-effect relationships to make evidence-based claims about science content, namely how pushes and pulls (forces) influence the motion of an object. This process of exploration allows students to engage in scientific inquiry, developing their skills in observation, data collection, and analysis. By interpreting the data they gather, students can construct a deeper understanding of scientific principles and articulate their findings in the form of evidence-based claims. This hands-on approach not only reinforces content knowledge but also fosters critical thinking and problem-solving skills.

How Are Assessments Incorporated?

Assessments are used seamlessly throughout the lesson and serve many purposes. Firstly, assessments like notice and wonder questioning routines create relevancy and activate student thinking at the lesson's onset. Teacher-student discourse asks students to consider the factors needed to ensure fair tests that produce valid and reliable results. In addition, when data is in hand, students are prompted to use patterns and cause-and-effect relationships and develop thinking versus introducing content and validating students' ideas. The conversations that occurred and revisiting initial ideas are also a formative assessment and promote students' thinking about developing understandings. Formative assessments are critical because they provide ongoing feedback to both students and teachers about the learning process. They help identify areas where students may need additional support or clarification, allowing for timely interventions that can enhance understanding. Summative assessments, on the other hand, evaluate student learning at the end of an instructional unit by comparing it against some standard or benchmark. Both types of assessments are essential in a well-rounded educational approach, ensuring that learning is both comprehensive and meaningful.

What Teacher Enhancements Are Needed to Develop a Deeper Understanding?

Teacher enhancements allow students to use scientifically accurate terminology in light of their firsthand experiences (for grades 3 and beyond, the term "force"). Just-in-time learning introduces concepts related to Newton's second law for secondary students. Introducing ideas through discussions is a rich experience because students internalize new terms and concepts related to their developing conceptual frameworks that they have constructed first-hand. This approach is supported by educational theories that advocate for scaffolded learning, where teachers provide the necessary support structures to help students progress in their understanding. As students become more proficient, these supports are gradually removed, encouraging independent learning and critical thinking. Teacher enhancements also involve differentiating instruction to meet the diverse needs of learners, ensuring that each student can access and engage with the content meaningfully.

How Is *Explore-Before-Explain* Being Used?

This four- to five-day lesson on helicopter fall time illustrates the importance of sequencing science instruction according to the essential features of active meaning-making. Firstly, students' ideas and experiences are targeted to a relevant occurrence—the factors influencing maple seed fall time. These ideas are further developed by pinpointing and making predictions about a specific factor—whether maple tree seed size affects flight time. Further engagement is encouraged by using the inquiry planning wheel and paper helicopters as a model to ensure accurate and valid results are gained when it is time to explore. Next, students explore specific ideas so they have data that serves as evidence for active meaning-making and constructing evidence-based claims. Following students' evidence-based claims, the teacher provides enhancement-type activities to sophisticate understanding and ensures targeted Standards are addressed. Enhancements occur as both explanations and further explorations to deepen student meaning-making. Throughout the lesson, students consider their developing understanding and how they use practices and crosscutting concepts to understand science better.

The *Explore-Before-Explain* model aligns with research on effective teaching strategies, which suggests that students learn best when they have the opportunity to explore concepts and develop their understanding before being given formal explanations. This approach encourages active learning: students are participants in the learning process rather than passive recipients of information. It fosters a deeper understanding of the material, because students are more likely to remember and apply what they have learned when they have discovered the concepts for themselves. Furthermore, this model promotes a growth mindset, in that students view challenges as opportunities to learn and grow rather than insurmountable obstacles. By adopting an *Explore-Before-Explain* approach, teachers can create a more engaging and effective learning environment that encourages curiosity, critical thinking, and a lifelong love of learning.

Helicopter Fly Time Standards Connections and Progressions

The model lesson illustrates numerous Standards in practice. The Standards are seamlessly intertwined so that content (Disciplinary Core ideas [DCIs]) are learned by using Science and Engineering Practices (SEPS) and Crosscutting Concepts (CCC) (NGSS Lead States, 2013).

Conceptual Understandings (DCIs)

- Pushes and pulls can have different strengths and directions (K-2: PS).
- A situation we want to change or create can be approached as a problem to be solved through engineering. Such problems may have many acceptable solutions (K-2: PS).
- Each force acts on one particular object and has both strength and direction. An object at rest typically has multiple forces acting on it, but they sum to zero net force on the object. Forces that do not sum to zero can cause changes in the object's speed or direction of motion (3–5: PS).
- The motion of an object is determined by the sum of the forces acting on it; if the total force on the object is not zero, its motion will change. The greater the object's mass, the greater the force needed to achieve the same change in motion. For any given object, a more significant force causes a more considerable change in motion (6–8: PS).
- Newton's second law accurately predicts changes in the motion of macroscopic objects (9–12: PS).

Science and Engineering Practices

- Analyzing and Interpreting Data
- Developing and Using Models
- Planning and Carrying Out Investigations
- Constructing Explanations
- Obtaining, Evaluating, and Communicating Information

Crosscutting Concepts

- Patterns
- Model-Based Reasoning

Reference

NGSS Lead States. (2013). *Next generation science standards: For states, by states.* The National Academies Press. https://doi.org/10.17226/18290

Module 7

Applying Active Meaning-Making and *Explore-Before-Explain* to Practice

> **Purpose:** To focus our *Explore-Before-Explain* planning using time-efficient strategies.
>
> **Desired Results**
>
> Curriculum designers and teacher leaders will understand that:
>
> ★ Shifting focus requires clarity about learning goals.
> ★ Priority planning is about balancing what works in our current practice and shoring up other key elements of effective instructional design.
>
> **Module Design Goals:** In this module, you will learn how to priority plan *Explore-Before-Explain* lessons that seamlessly translate into the three dimensions of the *Frameworks*.
>
> **You should work on Module 7 if** you are developing a homegrown curriculum and want to design more impactful lessons.

The most notable success occurs in classrooms, schools, and districts where teachers build their curriculum. *Give yourself grace through the process*. A good homegrown curriculum respects the teacher's time and expertise but acknowledges that we should use what we know about how students learn best in our instructional practices.

At this point, most teachers' heads are spinning with ideas for how to make higher levels of active meaning-making through *Explore-Before-Explain* a reality for students. Fullan (2016) suggested, "Careful attention to a small number of key details during the change process can result in the experience of success, new commitments, and the excitement and energizing satisfaction of accomplishing something important" (p. 8) We can use four active meaning-making features heavily supported by research to rethink current instructional design and practice.

Step 1: Plan "Backward" (Constructing Claims) in Practice

If this process is new to you, a logical and often easy place to begin is by considering the hands-on activities you already use with students. Hands-on activities, often referred to as experiential learning opportunities, are critical in fostering a deeper understanding of scientific concepts. Start by compiling a comprehensive list of these activities (see Table 8.1 for an example). This list will serve as your foundation for identifying the evidence-based claims that students could potentially make from each activity.

Once you have your list, take the time to analyze each activity in detail. Consider the specific data that students gather during these activities. Data can come in various forms, such as measurements, observations, or experimental results. From this data, identify potential evidence-based claims that students might make. For instance, if an activity involves measuring plant growth under different light conditions, students might claim that plants grow faster under certain types of light based on their collected data.

You might find that articulating these evidence-based claims is not as straightforward as it initially seems. It's common to encounter challenges when trying to connect data with specific claims. The process of transforming raw data into coherent, evidence-based claims requires careful consideration of how the data supports or refutes a hypothesis. Reflect on the patterns you observe in the data or the cause-and-effect relationships that emerge. These patterns and relationships are crucial for forming valid scientific claims and developing a deeper understanding of the underlying principles.

It's important to recognize that if you encounter difficulties in determining the evidence-based claims, students are likely to face similar challenges. This underscores the importance of thoroughly understanding the connection between hands-on activities and the claims that can be derived from them. Providing clear guidance and support to students as they navigate this process is essential for their success.

Additionally, ensure that the hands-on, minds-on activities and the evidence-based claims students are encouraged to make are aligned with the content standards. The content standards, often outlined by educational

Table 8.1 Student evidence-based claims standards checklist

Content Area	Hands-on/Minds-on Experience	Students' Evidence-Based Claims	Standard Check (Disciplinary Core Ideas; NGSS Lead States, 2013)
Forces and Interactions	Exploring different materials to see if they are magnetic and whether magnets work in air and water (Brown & Keeley, 2023)	Only certain materials are magnetic; magnets do not have to touch other materials to work.	**Grades 3–5:** Types of Interactions: Electric, magnetic, and gravitational forces between a pair of objects do not require that the objects be in contact—for example, magnets push or pull at a distance. The sizes of the forces in each situation depend on the properties of the objects and their distances apart and, for forces between two magnets, on their orientation relative to each other.
Forces and Interactions	Exploring how differently shaped marble tracks influence motion (Brown & Keeley, 2023)	The shape of the track directs the forces that influence the speed and direction marbles travel.	**Grades K–2:** Pushing or pulling on an object can change the speed or direction of its motion and start or stop it. **Grades 3–5:** The patterns of an object's motion in various situations can be observed and measured; when past motion exhibits a regular pattern, future motion can be predicted from it.

(*Continued*)

Table 8.1 (Continued)

Content Area	Hands-on/Minds-on Experience	Students' Evidence-Based Claims	Standard Check (Disciplinary Core Ideas; NGSS Lead States, 2013)
Ecosystems	Exploring the needs of sprouting radish seeds (Brown & Keeley, 2023)	Sprouting seeds need water and warmth to grow but not sunlight or soil	**Grades K–2:** Germination, life cycles, needs of organisms, seeds; plants need to take in water and light to grow. **Grades 3–5:** Plants have external structures that support survival, growth, and reproduction.
Light Waves	Exploring the factors that influence shadow size and shape (Brown & Keeley, 2023)	The position and direction of the light source on the object impact the shadow created.	**Grades K–2:** Some materials allow light to pass through them, others allow only some light through, and others block all the light and create a dark shadow on any surface beyond them, where the light cannot reach.

(*Continued*)

Table 8.1 (Continued)

Content Area	Hands-on/Minds-on Experience	Students' Evidence-Based Claims	Standard Check (Disciplinary Core Ideas; NGSS Lead States, 2013)
Waves	Exploring water waves in a container to move a bobber side-to-side (Teacher note: not moving the whole container side-to-side) (Brown & Keeley, 2023)	A bobber will move up and down but not side to side when waves are made in a water container using force. The height and number of times the bobber goes up and down depends on the force or speed at which the water is moved and creates different shapes of waves.	**Grades 3–5:** Waves, which are regular patterns of motion, can be made in water by disturbing the surface. When waves move across the surface of deep water, the water goes up and down in place; it does not move in the direction of the wave—observe, for example, a bobbing cork or seabird—except when the water meets the beach.
Energy	Exploring how to make a simple series circuit (Brown & Keeley, 2023)	A battery must be connected in a complete loop, from one terminal to the bulb using wires and then back from the bulb's screw threads to the other battery terminal.	**Grades 3–5:** Energy can be transferred from place to place by sound, light, heat, and electrical currents. The energy transferred in a circuit can produce sound, heat, or light. **Grades 3–5:** Conservation of Energy and Energy Transfer: Energy can also be transferred from place to place by electric currents, which can then be used locally to produce motion, sound, heat, or light. The currents may have been produced by transforming the energy of motion into electrical energy.

(Continued)

Table 8.1 (Continued)

Content Area	Hands-on/Minds-on Experience	Students' Evidence-Based Claims	Standard Check (Disciplinary Core Ideas; NGSS Lead States, 2013)
Matter and Its Interactions	Exploring what happens when mixing baking soda and vinegar (Brown & Keeley, 2023)	Materials change and gas is produced, but the weight stays the same.	**Grades 3–5:** When two or more different substances are mixed, a new substance with different properties may be formed. **Grades 3–5:** The amount (weight) of matter is conserved when it changes form, even in transitions in which it seems to vanish.
Light Waves	Exploring whether materials reflect light (Brown & Keeley, 2023)	Regardless of whether a material is shiny or dull, if light hits it and we see the material, then it reflects light. We can only see objects in complete darkness if they emit light.	**Grades 3–5:** An object can be seen when light reflected from its surface enters the eye.
Earth's Systems	Exploring whether earth's materials can be moved with water (Brown & Keeley, 2023)	Water can move and break up some earth materials quickly, creating new and different surfaces.	**Grades 3–5:** Earth Materials and Systems: Rainfall helps to shape the land and affects the types of living things found in a region. Water, ice, wind, living organisms, and gravity break rocks, soils, and sediments into smaller particles and move them around.

(Continued)

Table 8.1 (Continued)

Content Area	Hands-on/Minds-on Experience	Students' Evidence-Based Claims	Standard Check (Disciplinary Core Ideas; NGSS Lead States, 2013)
Ecosystems	Exploring 2-liter bottle ecosystems (Brown, 2021; Brown & Concannon, 2018)	Both living and non-living parts of an ecosystem are connected, and living organisms depend on different parts of the environment for survival.	**Grades 3–5:** The food of almost any kind of animal can be traced back to plants. Organisms are related in food webs in which some animals eat plants for food and others eat the animals that eat plants. Some organisms, such as fungi and bacteria, break down dead organisms (plants or plants and animals) and therefore operate as "decomposers." Decomposition eventually restores (recycles) some materials back to the soil. Organisms can survive only in environments in which their particular needs are met. A healthy ecosystem is one in which multiple species of different types are each able to meet their needs in a relatively stable web of life. Newly introduced species can damage the balance of an ecosystem. **Grades 3–5:** Matter cycles between the air and soil and among plants, animals, and microbes as these organisms live and die. Organisms obtain gases and water from the environment and release waste matter (gas, liquid, or solid) back into the environment.

(Continued)

Table 8.1 (Continued)

Content Area	Hands-on/Minds-on Experience	Students' Evidence-Based Claims	Standard Check (Disciplinary Core Ideas; NGSS Lead States, 2013)
Light Waves	Exploring how different light travels through different materials (Brown, 2018)	The properties of materials can influence what we see and whether light bounces off the object or goes through the material.	**Grades 6–8:** When light shines on an object, it is reflected, absorbed, or transmitted through it depending on the object's material and the frequency (color) of the light. **Grades 6–8:** The path that light travels can be traced as a straight line except at surfaces between different transparent materials (e.g., air and water, air and glass) where the light path bends.
Energy	Exploring whether common materials are magnetic and whether magnets work at a distance (Brown, 2018)	Only certain metals are magnetic, and the magnet type and distance between magnets influence the force at which they push or pull other magnets.	**Grades 6–8:** Electric and magnetic (electromagnetic) forces can be attractive or repellent, and their sizes depend on the magnitudes of the charges, currents, or magnetic strengths involved and on the distances between the interacting objects.

(Continued)

Table 8.1 (Continued)

Content Area	Hands-on/Minds-on Experience	Students' Evidence-Based Claims	Standard Check (Disciplinary Core Ideas; NGSS Lead States, 2013)
Forces and Interactions	Exploring how the type of paper airplane influences flight distance (Brown & Concannon, 2019; Brown, 2018)	The forces acting on different styles of paper airplanes are different and influence flight distance.	**Grades 6–8:** For any pair of interacting objects, the force exerted by the first object on the second object is equal in strength to the force that the second object exerts on the first but in the opposite direction (Newton's third law).
Energy	Exploring how thermal energy transfers in a closed system (Brown, 2021)	Thermal energy transfers from hot to cold until all molecules reach the same (average) temperature.	**Grades 9–12:** Develop and use models to illustrate that energy at the macroscopic scale can be accounted for as a combination of energy associated with the motions of particles (objects) and energy associated with the relative positions of particles (objects). **Grades 9–12:** Plan and conduct an investigation to provide evidence that the transfer of thermal energy when two components of different temperatures are combined within a closed system results in a more uniform energy distribution among the components in the system (second law of thermodynamics).

(Continued)

Table 8.1 (Continued)

Content Area	Hands-on/Minds-on Experience	Students' Evidence-Based Claims	Standard Check (Disciplinary Core Ideas; NGSS Lead States, 2013)
Light Waves	Exploring how different materials influence the speed that light travels (Brown, 2021)	As light travels through different materials, the speed and angle can change.	**Grades 9–12:** The wavelength and frequency of a wave are related to one another by the speed of travel of the wave, which depends on the type of wave and the medium through which it is passing.
Forces and Interactions	Exploring the factors that influence how fast a car travels on a ramp (Brown, 2021)	Adding mass to a car but keeping the force constant results in a slower-moving vehicle. Increasing the force applied to move a car when the mass is constant results in a faster-moving car. Vehicles with different masses traveling down a sloped ramp move at the same speed.	**Grades 9–12:** Newton's second law accurately predicts changes in the motion of macroscopic objects.

(Continued)

Table 8.1 (Continued)

Content Area	Hands-on/Minds-on Experience	Students' Evidence-Based Claims	Standard Check (Disciplinary Core Ideas; NGSS Lead States, 2013)
Matter and Its Interactions	Exploring the results of a chemical reaction (Brown, 2021)	When baking soda and vinegar are mixed, the mass and the types of atoms are conserved. However, the arrangement of atoms is different for the reactants and products.	**Grades 9–12:** The fact that atoms are conserved, and knowledge of the chemical properties of the elements involved, can be used to describe and predict chemical reactions.

frameworks or curriculum guidelines, set clear goals for student learning. By cross-referencing your activities and claims with these standards, you ensure that your instructional practices remain focused on the intended learning outcomes.

If you work in grade-level teams or independently, create detailed tables of your current hands-on activities. These tables should not only list the activities but also include columns for the type of data collected, possible evidence-based claims, and how these claims align with content standards. For example, you could use tables similar to those found in *Instructional Sequence Matters* (Brown, 2019) and *Activating Students' Ideas: Linking Formative Assessment Probes to Instructional Sequence* by Brown and Keeley (2023). These resources provide examples and frameworks that can guide you in developing your own tables.

You might wonder why the focus is primarily on content from the standards. The reason is that students can only construct evidence-based claims by integrating a unique combination of science and engineering practices and crosscutting concepts. These practices and concepts are integral to the standards and provide the foundation for students to make valid scientific claims. By emphasizing the alignment with standards, you ensure that your activities and claims are grounded in the essential principles of science education.

In summary, expanding your understanding of how hands-on activities can lead to evidence-based claims involves careful analysis of the data, alignment with educational standards, and creating supportive tools such as detailed tables. By taking these steps, you enhance your ability to facilitate meaningful scientific inquiry and ensure that your instructional practices effectively support student learning.

As a word of caution, while it may be tempting to dive right into creating an *Explore-Before-Explain* lesson using the first topic in your scope and sequence (or, if you are reading this during the school year, to select the next topic in your sequence), it is often more practical to start with activities that you have already found to be effective. Begin with tried-and-true activities from your hands-on experiences evidence-based claims Standards checklist. This approach allows you to leverage your existing experience and insights, which can make the process more manageable and less overwhelming.

Starting with familiar activities offers several advantages. Firstly, it provides a solid foundation on which you can build your *Explore-Before-Explain* lessons. Since these activities have been successful in the past, you can more easily focus on refining your approach and enhancing your lesson plans rather than grappling with the uncertainties of new activities. This way, you can concentrate on developing your skills in designing and implementing *Explore-Before-Explain* lessons without the additional challenge of testing untried activities.

Additionally, using tried-and-true activities gives you the opportunity to reflect on what has worked well in the past and identify any potential areas for improvement. By analyzing these activities, you can gain a deeper understanding of how students interact with the content and how their evidence-based claims can be better aligned with the Standards. This reflection can provide valuable insights that will inform your approach to new and more complex *Explore-Before-Explain lessons*.

Starting with familiar activities also ensures that you have ample time to think through and develop more intricate *Explore-Before-Explain* lessons that may not be as straightforward. This approach allows you to build confidence and proficiency before tackling more challenging topics. By targeting evidence-based claims that students can make from your hands-on, minds-on experiences and ensuring these claims are closely tied to the Standards, you set a strong foundation for successful instructional planning.

In summary, while it may be tempting to leap directly into new topics, beginning with activities you know well can streamline your process and provide a clearer path for developing effective *Explore-Before-Explain* lessons. This strategy not only leverages your existing expertise but also offers a chance to hone your skills and ensure alignment with educational standards.

Step 1 Content: Consider an example by picking one topic from Table 8.1. For example, many teachers across the K-12 spectrum mix baking soda and vinegar only to realize students do not understand whether materials and mass change during a chemical reaction—a fundamental idea in the Standards (Driver et al., 1994).

Content Area	Hands-on/Minds-on Experience	Students' Evidence-Based Claims
Matter and Its Interactions	Exploring what happens when mixing baking soda and vinegar	Materials change and gas is produced, but the weight stays the same.

Standard Check (Disciplinary Core Ideas; NGSS Lead States, 2013)

Grades K–2

★ Different states of matter exist (e.g., solids, liquids, gases).
★ Objects or samples of a substance can be weighed, and their size can be described and measured.

Grades 3–5

★ The amount (weight) of matter is conserved when it changes form, even in transitions in which it seems to vanish.
★ No matter what reaction or change in properties occurs, the total weight of the substances does not change. (Boundary: Mass and weight are not distinguished at this grade level.)

Grades 6–8

★ Substances react chemically in characteristic ways. In a chemical process, the atoms that make up the original substances are regrouped into different molecules, and these new substances have different properties from those of the reactants.
★ The total number of each type of atom is conserved; thus, the mass does not change.

Grades 9–12

★ The fact that atoms are conserved, together with knowledge of the chemical properties of the elements involved, can be used to describe and predict chemical reactions.
★ Chemical processes, their rates, and whether or not energy is stored or released can be understood in terms of the collisions of molecules and the rearrangements of atoms into new molecules.

Step 1 Example

Exploring a Chemical Change: *www.youtube.com/watch?v=gIuJKE6eGpQ*

Students can make evidence-based claims about the products and reactants in a chemical change and conservation of matter from a purposeful science exploration. To conduct the exploration shown in the video resource, first place 50 ml of vinegar in an Erlenmeyer flask. Next, place baking soda in a balloon. Carefully stretch the balloon over the mouth of the Erlenmeyer flask, making sure not to mix the baking soda and vinegar. Add a small strip of black electrical tape to seal the balloon to the Erlenmeyer flask with no gaps. Zero out the electronic scale and then place the flask-balloon setup on a balance. Finally, lift the balloon, letting the baking soda come into contact with the vinegar. When teachers purposefully design lessons using backward design, students can construct evidence-based claims that materials change (gas production as evidenced by blown-up balloon) and that weight (mass) is conserved.

Materials Needed for the Activity

- Baking soda
- Vinegar
- Balloon
- Small Erlenmeyer flask (50 ml) or clear glass bottle
- Electronic balance or digital kitchen scale
- Indirectly vented chemical splash goggles
- Nitrile gloves
- Non-latex aprons

Safety Notes

1. Make sure you stand at a safer distance from the teacher's demonstration—15 feet or more.
2. The demonstrator will wear sanitized indirectly vented chemical-splash safety goggles meeting ANSI/ISEA Z87.1 D3 standard, nitrile gloves, and a non-latex apron during the activity's setup, hands-on, and takedown segments.
3. Observers will wear indirectly vented chemical-splash safety goggles during the demonstration's setup, hands-on, and takedown segments.
4. Quickly wipe up spilled or splashed water off the floor so it does not become a slip-and-fall hazard.
5. Use caution when handling glass, which can shatter if dropped and cut or puncture skin.
6. An eyewash station is required in case of a splash with hazardous chemicals or contact with physical hazards.
7. Hard printed or electronic copies of safety data sheets for all hazardous chemicals used are required.
8. Wash your hands with soap and water after completing this activity.

Step 2: Engaging Students' Ideas in Practice

Beginning new lessons with students' ideas and experiences creates a compelling "need-to-know" situation, which effectively sets the stage for meaningful explorations. By connecting lessons to students' own ideas and experiences, you engage their curiosity and motivate them to explore the content more deeply. This approach not only piques their interest but also fosters a sense of relevance and personal connection to the material.

To effectively activate student thinking, consider addressing the following two instructional design questions.

How Will You Elicit Students' Ideas and Experiences?

This question is crucial because it determines how you will tap into students' existing knowledge and perspectives. Eliciting students' ideas involves asking thought-provoking questions, encouraging discussion, and incorporating activities that prompt students to share their personal experiences. For instance, you might start a lesson by asking students to discuss a relevant real-life situation or conduct a brief survey to gather their initial thoughts on a topic. By doing so, you not only validate their experiences but also create a foundation upon which new learning can build.

What Conceptual Idea, Occurrence, or Event Will You Use to Anchor Learning?

In other words, how will you introduce the exploration in a way that makes it meaningful and relevant to students' lives? Choosing an appropriate anchor involves selecting a concept, event, or scenario that resonates with students and highlights the importance of the lesson. For example, if you're teaching about ecosystems, you might start with a current news event related to environmental changes or a local ecological issue. This approach helps students see the practical implications of their learning and establishes a context that enhances their engagement and understanding.

Both questions are integral to activating student thinking right from the lesson's onset. By effectively eliciting students' ideas and anchoring the learning in relevant contexts, you ensure that the exploration is not only engaging but also deeply connected to students' lives and experiences. This approach fosters a more interactive and dynamic learning environment, in which students are more likely to be invested in their learning journey.

> **Step 2 Example:** Suppose we follow through with exploring a chemical change example from earlier. In that case, teachers can elicit student ideas and experiences before the chemical reaction demonstration and at the onset of the unit. Using a notice and wonder routine, teachers can ask for observations of the chemicals (i.e., vinegar and baking soda) and curiosities students might have when the chemicals are mixed. For students who have seen the chemicals combined in early grades, teachers can ask for a rule for their thinking or evidence from past experiences. Teachers can also prompt thinking about common chemical and physical changes in students' lives.

Students can consider what happens when they bake a cake. Consider asking students about their ideas about whether *"Do you think a cake be unbaked to return the initial ingredients?"* Teachers can also use content-based formative assessment probes to ask whether students think materials change (e.g., *What is the Result of Chemical Change*, Keeley & Cooper, 2019) and for secondary students if the types and number of atoms are the same before and after a chemical reaction (e.g., *What Happens to Atoms During a Chemical Reaction*, Keeley & Cooper, 2019). Beginning new lessons with students' ideas and experiences activates learning and sets up explorations.

Student Ideas Alert!

The following are common student ideas identified in the research concerning chemical changes.

★ Andersson (1991) investigated children's notions of chemical change and found they appear to fall into six categories: (1) it just happens; (2) matter disappears; (3) the product materials must have been inside the starting materials; (4) the product material is just a modified form of the starting material; (5) the starting material just turns into the product material; and (6) the starting materials interact to form the product materials. Students experience difficulty discriminating consistently between a chemical change and a physical change. Evidence for this comes from several studies. For example, Ahtee and Varjola (1998) explored 13- to 20-year-olds' ideas about what kinds of things would indicate a chemical reaction had occurred. They found that about 20% of the 13- to 14- and 17- to 18-year-olds thought dissolving and change of state were chemical reactions. Only 14% of the 137 19- to 20-year-old university students in the study could explain what actually happened in a chemical reaction.

★ Vogelezang (1987) found that students who regard ice as a different substance from water are likely to consider freezing water or melting ice as a chemical change. Briggs and Holding (1986) reported that 75% of the secondary students they sampled thought a change in mass was evidence of a chemical change. Stavridou and Solomonidou (1989) explored ideas held by Greek students ages 8 to 17 by presenting them with 18 different phenomena to classify as a chemical or physical change. They found that students who used the reversibility criterion could distinguish between chemical and physical changes better than students who did not consider

> reversibility. The students who used the reversibility criterion considered chemical changes to be irreversible, which could pose a problem in understanding chemical reactions. Both groups used criteria that were macroscopic in character.
> ★ Abraham et al. (1994) found that students were confused about chemical and physical changes. There were indications that they had memorized the terminology rather than developed conceptual understanding.
> ★ In a study by Abraham et al. (1992), eighth-grade students observed a chemical change in which a glass rod was held in the flame of a burning candle and a black film formed on the rod. To show an understanding of chemical change, students were expected to identify the transformation and know that a new substance was formed, not just a different form of the same substance. Of the students questioned, some showed some understanding of chemical change. In contrast, others had some knowledge of chemical change but then provided evidence of a physical change, and some of them said the change was not a chemical change because no chemicals were involved. And 70% showed no understanding that a chemical change had occurred with the burning of the candle and the formation of the black film on the glass rod.

Step 3: Enhance Understanding (Connecting Claims to Scientific Principles) in Practice

Enhancement activities come in many forms and serve a variety of purposes, each designed to deepen and enrich student learning. These activities are crucial for extending the learning experience beyond the initial hands-on explorations. To effectively develop and implement enhancement activities, it is essential to use the Standards as a guide. The Standards provide a framework for determining the next steps and ensuring that enhancements are aligned with the learning goals.

Given that there are only so many hours in the school day and an extensive amount of science content to cover, it is impractical to address every aspect of the curriculum solely through hands-on activities. This is where enhancement activities play a pivotal role. For example, using a teacher's explanation as an enhancement can introduce key terminologies and concepts that may not be readily accessible through hands-on experiences alone. This approach helps to bridge the gap between the exploratory activities and more abstract scientific concepts.

It is also important to ensure that enhancement activities address and answer essential questions that contribute to the development of deeper scientific understanding. These activities should effectively link explorations with explanations, helping students to make connections between what they have experienced and the broader scientific concepts being taught. When developing explanations, consider the following prompts.

What Key Terms, Concepts, and Ideas are Highlighted in the Standards?

Identifying these elements ensures that the explanations provided during enhancement activities are directly relevant to the Standards and address the critical content that students need to understand.

How Will Explanations be Emphasized as Essential Questions to be Answered for Students?

By framing explanations around essential questions, you help students focus on the most important aspects of the content and encourage them to think critically about the material. This approach ensures that explanations are not just supplementary but integral to the learning process.

Answering both of these questions is crucial for effectively navigating the limitations of hands-on experiences. By providing thoughtful and targeted explanations, teachers can enhance the learning experience and ensure that students gain a comprehensive understanding of the scientific concepts. Enhancement activities, therefore, play a key role in bridging explorations with explanations and maximizing the educational impact of the limited time available for science instruction.

> **Step 3 Example (Enhancement as a Discussion):** Using the example of exploring a chemical change, we can illustrate how enhancements can determine where to go next when confronted with limits to our understanding. An essential question that could drive the lesson is whether all changes are the same, focusing on the differences between physical and chemical changes (this topic targets 3–5 standards). This is a perfect place to discuss examples illustrating the difference between chemical and physical changes. For instance, chemical changes can create new substances with different properties and are usually irreversible. Chemical changes are often evidenced by gas production, changes in color, the creation of an odor, or the formation of a precipitate. Physical changes do not create new substances but rather alter their form or state. In addition, here is a good place to define a substance so students can use rules about physical and chemical changes to other examples. A substance is a specific type of matter with a constant composition and distinct properties.

> For secondary students, the discussion can focus on macro-microscopic connections and what students know about atoms and molecules during a chemical change or reaction.

A second form of enhancement is to provide elaboration activities, which involve offering further explorations for students to deepen their understanding. These elaboration activities should be closely tied to the Standards to ensure they support the intended learning outcomes. Unlike initial explorations, elaborations build upon what students have already learned, bridging existing knowledge with new experiences. This approach allows students to transfer their ideas to different situations, thereby enhancing their conceptual understanding and application of scientific principles.

Elaboration activities offer the added benefit of helping students to make connections between various concepts and scenarios, which strengthens their overall grasp of the content. By engaging in these activities, students can apply their knowledge in diverse contexts, reinforcing their learning and developing a more robust understanding of the Standards. Consider the following prompt when planning elaboration-type enhancements.

Are Different Hands-on, Minds-on Experiences Needed to Further Develop Students' Conceptual Understanding of Standards?

Reflecting on this question will help you determine whether additional activities are required to deepen students' grasp of key concepts and ensure that they can effectively apply their knowledge in new and varied situations.

The answer to this question is crucial for establishing the next steps needed to help students fully comprehend and integrate ideas presented in the Standards. By thoughtfully designing elaboration activities, you can provide students with opportunities to explore and solidify their understanding, thereby enhancing their overall learning experience.

> **Step 3 Example (Enhancement as Further Exploration):** Elementary (and secondary) students can be confronted with whether a change of state is a physical or chemical change using their *Explore-Before-Explain* experiences. For example, students should explore whether when the ice melts to a liquid or is heated and becomes a gas is an example of a chemical or physical change (following the same safety as before). Students can investigate whether a new and different substance is produced when ice melts to water and whether this change of state is reversible.

Secondary students must develop a deeper understanding of chemical changes that bridge the macro-to-microscopic level (NGSS Lead States, 2013). Using computer simulations such as PhET's balancing chemical equations (see *https://phet.colorado.edu/en/simulation/balancing-chemical-equations*) or using molecular modeling kits, students can connect their evidence-based claims about the conservation of mass to types and numbers of atoms being the same before and after a chemical reaction. In addition, students can have elaboration activities exploring the products and reactants in various simple chemical reactions to test the utility of their budding ideas.

Student Ideas Alert!

The following are common student ideas identified in the research.

- ★ In general, students have difficulty developing an adequate conception of the chemical combination of elements until they can interpret the combination at the molecular level (Driver et al., 1994).
- ★ Although in science, the term *chemical change* refers to processes in which the reacting chemical substances transform into new substances, several studies have found that students often use the term to encompass a wide variety of changes, including physical transformations, especially when the color of a substance changes. How well students distinguish between chemical and physical changes may depend on their conception of the term *substance* (Driver et al., 1994).
- ★ Ben-Zvi et al. (1982) found that students have great difficulty changing their thinking when they are asked to transition from observable changes in substances to the atomic-molecular level to explain observable changes in terms of the interactions between individual atoms and molecules.
- ★ To master chemistry, students must understand chemical ideas at three levels: macroscopic, particle, and symbolic (Gabel, 1999).

Step 4: Promoting Reflection on Learning in Practice

Assessment from an active meaning-making perspective is crucial for helping students to reflect on their developing understanding. This approach emphasizes the importance of ongoing evaluation and encourages students to think critically about their learning process. Throughout the planned activities, teachers should take on the role of a guide who prompts students with probing questions such as, "How have your ideas changed and why?" These questions are designed to facilitate deep reflection and encourage students to articulate how their understanding has evolved.

At the end of the lesson, it is important for students to revisit their initial responses to engagement questions and provide new scientific claims based on the data they have gathered from their firsthand experiences. This reflective process allows students to connect their earlier ideas with the new knowledge they have acquired, demonstrating their growth and comprehension of the scientific concepts.

Probing and reflection questions are instrumental in helping students to think critically about their developing understanding. They encourage students to solve new problems, address unresolved questions, and consider how their experiences have influenced their current knowledge. Such questions also guide students in evaluating their learning and understanding how their experiences have either enhanced or constrained their grasp of the material.

Incorporating questioning prompts throughout the lesson is an effective strategy. These prompts should encourage students to accurately self-assess their learning, reflect on their experiences, and consider how these experiences have impacted their understanding. Additionally, allowing students to set future learning goals based on their reflections helps them take ownership of their learning journey and fosters a growth mindset. By integrating these elements into your assessment strategy, you enhance the overall learning experience and support students in achieving a deeper and more meaningful understanding of the scientific content.

Step 4 Example: While student-visible learning (i.e., reflection) should be encouraged throughout, the end of the lesson provides opportune times for students to think about their learning. At the end of the lesson, students should revisit their initial answers to pre-assessment items and offer new scientific claims based on data from their firsthand experiences about whether materials and weight (mass) change during a chemical reaction. Probing and reflection questions that ask students to think about their developing understanding help them solve new problems and answer questions for future scientific questions.

Using these four active meaning-making steps in instructional planning helps translate research into effective practice and centers lessons on the most salient features of how students learn best. By incorporating these steps, you ensure that your instructional design is rooted in research-backed strategies, which enhances the overall learning experience. Understanding these elements allows for a transformative shift in your approach to instructional design, enabling students to engage in the intellectual work necessary for deep learning.

The four steps, described next, can serve as the core process for rethinking and revitalizing instructional design.

Explore-Before-Explain in Practice

This approach integrates the four salient features of active meaning-making into a coherent instructional sequence. Initially, you begin by eliciting students' prior experiences and ideas related to a common science concept, event, or occurrence. This step sets the stage for meaningful exploration. Following this, students engage in firsthand experiences that enable them to construct evidence-based claims. This hands-on involvement is crucial for developing a deeper understanding of the concepts. Subsequently, enhancements are introduced to help students achieve or repeat the Standards-based goals, further solidifying their learning. Finally, students reflect on their developing understanding and learning process, engaging in authentic scientific inquiry.

Active Meaning-Making Through Explore-Before-Explain Template in Practice

The principles discussed can be organized into a practical template for instructional design that honors the need for students to actively construct meaning and emphasizes the significance of prior knowledge and experiences. This template is designed to streamline the instructional planning process, making it easier for educators to integrate these principles into their teaching practices. By structuring the planning steps into an *Explore-Before-Explain* sequence, the template simplifies the process and facilitates realistic and manageable changes to instructional design. This method not only supports a more effective learning experience but also helps teachers implement evidence-based practices in a systematic and organized manner.

In summary, utilizing these four active meaning-making steps and incorporating them into an *Explore-Before-Explain* instructional sequence enhances the quality of education by aligning teaching methods with how students learn most effectively. The template provided serves as a valuable tool for teachers, making it easier to apply these principles in practice and ensuring that instructional design is both impactful and responsive to student needs (see Table 8.2).

Table 8.2 Planning, essential elements, and activity example template

Planning Step	Active Meaning-Making Elements	Activity Examples
1	Targeting Evidence-Based Claims	• Current Hands-On Activity → Students' Evidence-Based Claim → Standards Check
2	Engaging Students' Ideas	• How will you elicit students' ideas and experiences? • What conceptual idea, event, or occurrence will you use to anchor learning? • How will you introduce the exploration so that it is meaningful and relevant to students' lives?
3	Enhancing Students Ideas	• What key terms, concepts, and ideas are needed that are highlighted in the Standards? • How will explanations be emphasized as essential questions to be answered for students? • Are different hands-on, mind-on experiences needed to develop students' conceptual understanding of standards further?
4	Reflecting on Learning Example	• How will students perform one or more of the following during the lesson? (1) Accurately self-assess. (2) Reflect on their learning. (3) Set future learning goals.
5	*Explore-Before-Explain* Sequence	
Step 2 → Step 1 → Step 3 → Step 4		

References

Abraham, M., Grzybowski, E., Renner, J., & Marek, E. (1992). Understandings and misunderstandings of eighth-graders of five chemistry concepts found in textbooks. *Journal of Research in Science Teaching*, 29(2), 105–120. https://doi.org/10.1002/tea.3660290203

Abraham, M., Williamson, V., & Westbrook, S. (1994). A cross-age study of the understanding of five chemistry concepts. *Journal of Research in Science Teaching*, 31(2), 147–165. https://doi.org/10.1002/tea.3660310207

Ahtee, M., & Varjola, I. (1998). Students' understanding of chemical reactions. *International Journal of Science Education, 20*(3), 305–316. https://doi.org/10.1080/0950069980200305

Andersson, B. (1991). Pupils' conception of matter and its transformations (age 12–16). *Studies in Science Education, 18*(1), 53–85. https://doi.org/10.1080/03057269108559956

Ben-Zvi, R., Eylon, B., & Silberstein, J. (1982). *Students vs. chemistry: A study of student conceptions of structure and process* [Paper presentation]. Annual Conference of the National Association for Research in Science Teaching, Fontana, WI.

Briggs, H., & Holding, B. (1986). *Aspects of secondary students' understanding of elementary ideas in chemistry.* Children's Learning in Science Project, University of Leeds.

Brown, P., & Keeley, P. (2023). *Activating students' ideas: Linking formative assessment probes to instructional sequence.* NSTA Press.

Brown, P. L. (2018). *Instructional sequence matters, grades 6–8: Structuring lessons with the NGSS in mind.* NSTA Press.

Brown, P. L. (2019). *Instructional sequence matters: Explore-before-explain, grades 3–5.* NSTA Press.

Brown, P. L. (2021). *Instructional sequence matters: Explore-before-explain physical science, grades 9–12.* NSTA Press.

Brown, P. L., & Concannon, J. (2018). *Inquiry-based science activities in grades 6–12.* Routledge.

Brown, P. L., & Concannon, J. (2019). *Evidenced-based science activities in grades 3–5: Meeting the NGSS.* Routledge.

Driver, R., Squires, A., Rushworth, P., & Wood-Robinson, V. (1994). *Making sense of secondary science: Research into children's ideas.* Routledge.

Fullan, M. (2016). *The new meaning of educational change* (5th ed.). Teachers College Press.

Gabel, D. (1999). Improving teaching and learning through chemistry education research: A look to the future. *Journal of Chemical Education, 76*(4), 548–554. https://doi.org/10.1021/ed076p548

Keeley, P., & Cooper, S. (2019). *Uncovering student ideas in physical science, Volume 3: 32 new matter and energy formative assessment probes.* NSTA Press.

NGSS Lead States. (2013). *Next generation science standards: For states, by states.* National Academies Press. www.nextgenscience.org/next-generation-science-standards

Stavridou, H., & Solomonidou, C. (1989). Physical phenomena–chemical phenomena: Do pupils make the distinction? *International Journal of Science Education, 11*(1), 83–92. https://doi.org/10.1080/0950069890110106

Vogelezang, M. (1987). Development of the concept of "chemical substance": Some thoughts and arguments. *International Journal of Science Education, 9*(5), 519–528. https://doi.org/10.1080/0950069870090506

Module 8

A New Mindset to Teaching

Purpose: To motivate teachers for the leadership and change possibilities that exist through supportive and purposeful environments.

Desired Results

Curriculum designers and teacher leaders will understand that:
★ Positive change can happen quickly if teachers have knowledge, skills, and abilities to make reform a reality in their teaching environments.

Module Design Goals: This module cultivates a growth mindset and the powerful influence one teacher, one leader, one team, once class/section can have on widespread change.

You should work on Module 8 if you wish to motivate teachers about their remarkable abilities to influence educational reform.

At this point, you have had various experiences (engaging in model lessons, learning about emerging research, having concrete instructional design principles, and using teachers' experiences as assets for learning). All of these ideas have been purposefully orchestrated to promote fundamental changes in how learning is viewed. Even with the experiences provided so far, sometimes, change can be difficult because it is not the norm of our school culture and district practices. Promoting change is problematic for many leaders because it differs from our colleagues' memories of how

> we learned science. Remember, research shows that mindsets are tenacious and can facilitate or hinder the changes we want to make.
>
> I use this example of outside education to illustrate that it can take just one person to transform the practices in a grade level, school, or district. Remember, do not underestimate the power of your influence on others and systems of practice. Mindsets are crucial considerations. They can facilitate and hinder the changes we want for our teams and students.

While the idea of mindsets might be new to you, they have a powerful impact on what we can accomplish. A mindset, in essence, is a set of beliefs or a mental attitude that determines how we interpret and respond to situations. When we adopt a growth mindset, we believe that our abilities and intelligence can be developed through dedication and hard work. This stands in contrast to a fixed mindset, in which we see our abilities as static and unchangeable. The power of a growth mindset is well illustrated by the dramatic change in paradigms that occurred in the running community after only 3 minutes and 59.4 seconds!

In 1954, Roger Bannister became the first person to break the 4-minute mile, a feat that was long considered impossible. For years, the 4-minute mile stood as a seemingly insurmountable barrier. Experts believed that the human body was not capable of achieving such a feat due to the limitations of the cardiovascular and respiratory systems. However, Bannister, who was both an accomplished runner and a medical student, believed otherwise. At the age of 25, he saw the 4-minute mile barrier not as an unbreakable limit but as a challenge that could be overcome with the right approach and mindset.

Bannister's journey to breaking the 4-minute mile was purposeful and methodical. His training regimen was innovative for its time and included several key components. Firstly, he incorporated shorter sprints that were run at a pace faster than his target mile pace. These high-intensity intervals helped him build the speed and endurance needed to maintain a fast pace over a longer distance. Secondly, Bannister developed a comprehensive endurance training program that allowed him to sustain his speed throughout the mile. Thirdly, he utilized "rabbits" or pacers—other runners who would set the pace for him during the race. These pacers were crucial in helping Bannister maintain the necessary speed without having to focus on setting the pace himself.

The day Bannister broke the 4-minute barrier was a momentous occasion. His time of 3:59.4 was not just a personal victory but also a groundbreaking moment in sports history. Bannister's achievement demonstrated the power of a growth mindset and the importance of believing in one's ability to achieve the seemingly impossible. His accomplishment was the result of consistent training, unwavering belief in his goal, and a strategic plan that aligned his knowledge, beliefs, and actions.

Bannister's record stood for only 46 days before it was broken again. This quick succession of breaking the 4-minute barrier by other runners highlights a critical aspect of mindset: once one person proves that a goal is achievable, the door is opened for others to believe and achieve the same. Since Bannister's historic run, more than 1,500 runners, including high school athletes and individuals over the age of 40, have broken the 4-minute mile. The current world record for the mile, held by Hicham El Guerrouj, stands at an astounding 3:43.13, set in 1999. This progression shows how a shift in mindset can lead to continuous improvement and achievement beyond previously imagined limits.

The "Bannister effect" resonates far beyond the world of sports. It underscores the importance of mindset in all areas of life, including education. As educators, we can draw valuable lessons from Bannister's story. The belief that students can achieve great things, coupled with a strategic and supportive approach, can lead to remarkable outcomes. Just as Bannister rethought the conventional training methods to achieve his goal, educators can rethink traditional teaching methods to enhance student learning and engagement.

Consider how the ideas of the Bannister effect might influence your mindset and abilities in fostering widespread scientific understanding. In the classroom, adopting a growth mindset means believing that all students have the potential to learn and excel in science. This belief should be reflected in our teaching practices. For example, providing students with challenging but achievable tasks, offering constructive feedback, and encouraging a positive attitude towards learning and mistakes can help students to develop resilience and a love for learning.

Moreover, educators can create a learning environment that emphasizes the importance of effort and perseverance. Just as Bannister's consistent and dedicated training led to his success, students who are encouraged to persist in the face of challenges are more likely to develop a growth mindset. By celebrating efforts and progress, rather than just the end results, teachers can help students understand that their abilities can grow with time and practice.

Furthermore, the concept of using "rabbits" or pacers in Bannister's training can be translated into the educational context. In this sense, "rabbits" could be mentors, peer tutors, or collaborative group work by which students support and challenge each other to achieve their best. Creating a community of learners who motivate and inspire each other can significantly enhance the learning experience.

Your experiences as an educator are critical in promoting active meaning-making in your practice and leading your teams. Just as Bannister combined his knowledge of medicine with his athletic training, teachers can integrate their diverse experiences and expertise to innovate in the classroom. Reflecting on your teaching practices, being open to new ideas, and continuously seeking ways to improve can lead to significant advancements in how students learn and engage with science.

In conclusion, the story of Roger Bannister breaking the 4-minute mile is more than just a tale of athletic achievement; it is a powerful example of the impact of mindset on what we can accomplish. By believing in our potential, setting high but attainable goals, and developing a strategic plan, we can achieve remarkable things. This principle applies not only to sports but also to education and many other fields. Embracing a growth mindset, both as educators and in our students, can lead to breakthroughs in learning and personal development. The Bannister effect reminds us that once we believe something is possible, we open the door to limitless possibilities.

Module 9

Leading Change

Purpose: To review the powerful influence of beliefs on practice and the factors that may facilitate and constrain the development of teacher mindsets.

Desired Results

Curriculum designers and teacher leaders will understand:

★ The important role that beliefs play in the development of teacher knowledge and practice
★ Understanding of why change might be hard for some teachers
★ A measure to determine whether teacher's beliefs have changed as a result of the previous modules in this guidebook (Elevator Speech)

Module Design Goals: In this module, you will learn about the factors that facilitate and constrain teachers' mindsets and their powerful impact on knowledge and practice.

You should work on Module 9 if you work with teams that are skeptical that students can do hard intellectual work before teacher explanations OR for teacher teams whose hands-on experiences fall short of achieving desired conceptual understanding OR for teacher teams.

You might Skip Module 9 if your teachers have embraced an *Explore-Before-Explain* mindset.

Until this point, I hope there is compelling evidence about instructional design practices that lead to higher levels of learning and more motivated students. The instructional design practices are supported by research and emphasized in the model lessons. However, you may be tasked with preparing colleagues with less active meaning-making knowledge and *Explore-Before-Explain* experiences. Remember, a vital purpose of this guidebook is to promote teacher leadership and develop collective teacher efficacy. While the call is ambitious, knowledge of research closely connected to ideas for practice can help sway colleagues who may be weary.

Suppose your teams' background experiences were not heavily focused on science-content courses, and we teach several content areas (true of many elementary teachers). To cope with the challenges of science teaching, they use hands-on experiences to teach science. While students are engaged and motivated, their activities fail to develop deep conceptual understanding. It is possible to have hands-on, super-engaging activities that do not actually promote learning. These assertions are not meant to place blame. A typical research finding is that when teachers need more content knowledge but use hands-on activities with students, they may need to develop the conceptual understanding called for in the Standards (Appleton 2003). As shown in *Science Stories*, explanations are necessary so that students gain a deeper conceptual understanding and can use scientific vocabulary appropriately to represent concepts reinforced by the hands-on activity (Koch 2018). Primarily using hands-on activity without a focus on the desired conceptual knowledge goals is contrary to *Explore-Before-Explain*. Hands-on activity without a clear focus ignores the critical role that explanations, constructed by students and provided by teachers, play in developing scientifically literate students.

Suppose your teams' experiences were mostly traditional, meaning that the emphasis was on learning content through lectures. In the USA and many other countries, there is a culturally embedded script to how instruction is sequenced (in science and other areas). For example, the traditional sequence divides instruction into three phases: (1) inform, (2) verify, and (3) practice (Abraham & Renner, 1986). The *Inform-Verify-Practice* sequence places primacy on content introduced by teachers and allows students to verify these ideas through hands-on, minds-on, and practice-type activities. As shown in *Taking Science to School: Learning and Teaching in Grades K-8*, the stark reality is that many students come away from traditional science experiences viewing science as primarily the collection of facts. The *Inform-Verify-Practice* sequence is contrary to *Explore-Before-Explain* and neglects the critical role that students' prior experiences and ideas play and the influential role of explorations in constructing explanations. (See Side Bar: The Pitfalls of Lecture.)

> **Side Bar. *The Pitfalls of Lecture***
>
> A few studies paint a particularly insightful picture of the shortfalls of passive learning environments. A Twilight of the Lecture presents a meta-analysis of 225 studies on the effectiveness of instructional approaches in undergraduate courses compared to lectures and active learning in STEM classes. The findings showed that active learning increased students' grades and their ability to be successful in the course by 36%. Conversely, the predominant use of lectures increased student failure rates by 55%. Even some of the best lecturers have been shocked by what they have learned about student learning from more passive approaches. The experiences of Eric Mazur, Balkanski professor of physics at Harvard, showed that despite his consistently positive student evaluations and lofty goals for student learning, his lectures came up short in promoting deep conceptual understanding. Through a simple test of students' knowledge of forces in his physics course, he found that students had the same misconceptions at the end of the course as they had at the beginning and that his lectures had taught them "next to nothing."

While mindsets may be new, research consistently shows that belief systems play a dominant role in establishing the knowledge we develop and our professional practice (Brown et al., 2013). Whether your background experiences and current practices are primarily activity-based or content-driven through lectures, you can use aspects of these beliefs to shift towards an *Explore-Before-Explain* mindset.

An *Explore-Before-Explain* Mindset to Teaching

An *Explore-Before-Explain* mindset consists of beliefs supporting a student-centered learning environment while highlighting our essential role in their development. Student learning is not entirely dependent on the knowledge they construct through discoveries. Instead, they have chances to investigate and ask questions leading to further explorations. At some point, we leverage students' firsthand science experiences. Rather than have them construct all scientific knowledge independently, we provide guidance and resources to help them craft their explanations so that they are not starting from scratch. It would not make sense or be efficient for students to rediscover all knowledge, which took scientists centuries to accumulate. Students' ideas and skills are powerful forces that drive intellectual development. Equally fundamental is how we guide students to explanations of science in light of their life experiences.

Having an *Explore-Before-Explain* mindset means that in our planning, we prioritize giving students firsthand experiences with data, allowing students to construct evidence-based claims that focus on conceptual understanding, and challenging students to discuss and think about the *why* behind science events and occurrences. An *Explore-Before-Explain* mindset incorporates discovery, problem-based, and inquiry-based learning. These strategies are activity-dependent, and as *Explore-Before-Explain* teachers, we acknowledge that all these classroom activities should happen before we explain ideas and introduce academic vocabulary. Having an *Explore-Before-Explain* mindset affords a place for didactic instruction (i.e., lectures, readings) at a time when students' firsthand experiences conceptualize it. Their firsthand experiences provide a framework for developing a more sophisticated understanding.

Explore-Before-Explain teaching ensures that students' conceptual understanding is transferable and that students develop knowledge of facts, terms, additional skills, and critical thinking in light of their firsthand experiences. The explanations we guide students to (as well as the information provided in such things as technical readings, videos, and simulations) make students' understanding more sophisticated and allow for an accurate and coherent science story. The activities (labs, readings, discussions, simulations, lectures) and disciplines (science, English language arts, math) become tightly intertwined for learners. Students do not see these disciplines as different ways to know the world; instead, the focus is on creating a community of learners that uses the best tools and knowledge structures to develop understanding.

Reflection alert! #Eb4E Elevator Pitch: An elevator pitch is a short but persuasive speech to generate interest in your work. A good elevator pitch should be between 20 and 30 seconds, about the time it takes to ride in an elevator, hence the name. If I asked you right now whether you could give an elevator pitch about your science mindset, could you do it? If you struggle to think of what to say, consider your core beliefs about the purpose of science learning and your role as a science teacher.

★ What are your core beliefs about learners? List several ideas.
★ What does the best teaching environment look like? Generate a list of multiple ideas.
★ Analyze your list with a critical eye. Eliminate any redundancies or vague ideas. Home in on the ideas you feel are the best.
★ Write your ideas as a bulleted list. Revise your ideas if necessary, and organize them so that they are worded in a way that the general public will understand.
★ Organize your bulleted list in a logical order.

References

Abraham, M. R., & Renner, J. W. (1986). *The sequence of learning cycle activities in high school chemistry.* NARST Monograph, Number 1. National Association for Research in Science Teaching.

Appleton, K. (2003). How do beginning primary school teachers cope with science? Toward an understanding of science teaching practice. *Research in Science Education, 33*(1), 1–25. https://doi.org/10.1023/A:1023666618800

Brown, P. L., Abell, S. K., & Friedrichsen, P. (2013). The influence of conceptions of teaching and learning on a new teacher's developing practices. *Research in Science Education, 43*(2), 619–650. https://doi.org/10.1007/s11165-012-9306-x

Koch, J. (2018). *Science stories: Science methods for elementary and middle school teachers* (6th ed.). Cengage Learning.

Conclusions

If you embrace the ideas presented in this guidebook, you are part of a growing cohort of teachers devoted to promoting active meaning-making for students. Considering your mindset is essential because your beliefs about the roles of both student and teacher will filter how you make sense of and use the ideas in this guidebook. Adopting an *Explore-Before-Explain* mindset will give you new insights into instructional teaching practices that leverage the best possible learning experiences for students.

While mindsets may not shift overnight, gaining understanding by exploring research and the essential elements of active meaning-making in light of vivid examples, with ample opportunity to reflect, provides powerful professional learning. This gradual process ensures that the shift in mindset is deep and lasting. The mindset chapter came near the end to prepare teachers with a robust knowledge base grounded in classroom examples so their beliefs could shift through a practice-oriented approach. This strategic placement helps to ensure that the theoretical aspects of mindset are contextualized within practical experiences, making the transition smoother and more relatable for teachers.

This guidebook's ultimate goal is to go beyond changing a single teacher's classroom practice. While changing a single teacher's practice is a highly desirable intent, remember that we can have a broader impact by working together in teams, growing and learning from one another to develop higher levels of collective efficacy. Collective efficacy refers to the shared belief among a group of teachers in their ability to positively affect students. Research has shown that collective efficacy is a significant factor in improving student outcomes. By fostering a collaborative environment where teachers can share successes, challenges, and strategies, we create a supportive network that enhances individual and group effectiveness.

Let us review the learning targets to see whether we have addressed our targets for professional learning. Teachers were invited to actively participate and reflect on vivid examples of learning by active meaning-making through *Explore-Before-Explain* lessons tied to Standards. These lessons were carefully designed to illustrate the principles of the *Explore-Before-Explain* approach, which emphasizes student engagement and discovery before formal explanations are given. Three complete lessons—covering thermal energy transfer, the watermelon and grape experiment, and fall time—along with the design principles for teaching chemical changes and reactions were presented. Each of these lessons included reflection questions at key points so teachers could think about a particular phase of instruction. These questions were intended to prompt deep thinking and self-assessment, encouraging teachers to consider how the concepts and methods could be applied in their own classrooms.

Each activity was also purposefully placed in an *Explore-Before-Explain* sequence to create a storyline for learning driven by the gradual development of student conceptual understanding. The model lessons were not presented in a traditional format. Many resources present the research first and design principles prior to model lessons. My *Instructional Sequence Matters* series is written in this format—research first, followed by model lessons. This guidebook attempted to strategically place model lessons to 1) elicit teachers' ideas about sequencing instruction that may be new or different from their previous experiences and 2) introduce ideas, both research and active meaning-making design principles, in light of firsthand experiences with model lessons that illustrate these points.

By presenting model lessons in this manner, we aimed to challenge teachers to rethink their instructional approaches and consider the benefits of the *Explore-Before-Explain* method. This method aligns with how students naturally learn and process information, fostering a more engaging and effective learning environment.

A second learning target was to promote understanding of learning from a more research-based perspective. Overarching bodies of scholarship were chosen to highlight key ideas about learning and learners that have been extensively studied. These research findings provide a solid foundation for the instructional strategies discussed in the guidebook. The considerable overlap among research bodies only reaffirms the power of highlighting research in practice. Becoming an expert in one or more of these areas prepares teachers to be leaders who may be confronted with questions critical for understanding, "Why *Explore-Before-Explain* so important for learners?"

By grounding their practices in research, teachers can confidently advocate for innovative methods and address any skepticism from colleagues, administrators, or parents.

A third learning target was to connect key active meaning-making features with examples and explore them before explaining them in relation to contemporary research and the Standards. Each model lesson includes a Standards progression list of Disciplinary Core Ideas, Science and Engineering Practices, and Crosscutting Concepts. Many of the activities can be used across grade spans depending on the Standards targeted. The key in each lesson is that Standards drive the lesson's direction as the science story unfolds. Many teachers are shocked that they could use a similar activity at K-2 but also with 9–12 students. This flexibility demonstrates the universal applicability of the *Explore-Before-Explain* approach and its alignment with the Standards. The Standards pinpoint the conceptual understandings desired of students and the direction of needed enhancements through explanations and further exploration. The examples of active meaning-making were offered to provide teacher leaders with practical ways to develop curriculum in light of Standards and research. Teachers with a clear vision and focus can better use their knowledge to transform their instructional design practices into more powerful learning experiences for all students.

Finally, the question, "By what measures does the guidebook shift our professional practice?" is an assessment question. To answer this question, you could look at present and past student assessment scores and compare them to what is generally found in the research about learner-centered sequences of instruction (remember the research chapter in which *Explore-Before-Explain* students show higher achievement, motivation, and longer-lasting conceptual understanding than those who learn through traditional instruction). This comparative analysis can provide valuable insights into the effectiveness of the *Explore-Before-Explain* approach in your context. You could reflect on past and new data to see if these research elements hold true for your students and their classroom experiences. Additionally, consider qualitative measures such as student engagement, participation, and feedback to get a comprehensive view of the impact on learning.

I invite you to take an active meaning-making approach that honors the vital role that reflection plays in our developing understanding. Reflection is a critical component of professional growth and effective teaching. By regularly reflecting on your practices, student responses, and learning outcomes, you can continuously refine and improve your instructional methods. Consider your elevator pitch. "Does your elevator pitch include ties to research?" "Does the elevator pitch include the salient elements of active meaning-smaking and *Explore-Before-Explain*?" Finally, "Is your elevator pitch different after reading the guidebook from what it would have been before?" If your elevator pitch addresses these questions, you are on the path to becoming an *Explore-Before-Explain* leader devoted to promoting active meaning-making for students.

As a teacher leader, you have robust knowledge and convincing evidence to promote change in your class, grade span, building, district, and national levels. These changes for you and your team could usher in what Petrilli (2019) calls the "golden age" of science education reform because our students are "waiting." This "golden age" represents a time when innovative, research-based instructional practices are widely adopted, leading to significant improvements in student understanding, engagement, and achievement in science. By embracing and advocating for these methods, you can play a pivotal role in shaping the future of science education and ensuring that all students have the opportunity to develop a deep, meaningful understanding of scientific concepts.

In conclusion, the journey to adopting an *Explore-Before-Explain* mindset is ongoing and requires dedication, reflection, and collaboration. By engaging with the ideas and practices outlined in this guidebook, you are positioning yourself as a leader in educational innovation. The impact of your efforts will extend beyond your classroom, influencing colleagues and shaping the broader educational landscape. As you continue to explore, reflect, and implement these strategies, remember that your role as an educator is not just to impart knowledge but to inspire a love for learning and curiosity in your students. The transformation in your practice will not only benefit your students but also contribute to the collective efficacy of your teaching community, driving meaningful change in science education.

Reference

Petrilli, M. (2019). *Toward a golden age of educational practice*. Thomas B. Fordham Institute.

Index

Note: Page numbers in *italics* indicate figures, and page numbers in **bold** indicate tables in the text

5E instructional model 37, 38

Abraham, M. 109
activating students' ideas 66, 67, 73, 106, 107–108; activating thinking and meaning-making 67; bridging learning experiences 68; discrepant events 67; increasing student engagement 66–67; phenomena-based teaching 66; through prior knowledge 68; stimulating curiosity and inquiry 67
active learning 33, 68, 90, 123
active meaning-making 1, 62, 67, 128; contemporary standards 5–7; in deep science learning 24; high-quality professional learning 4; instructional core 3; integrating four steps into instructional design 114; leading effective science curriculum-based learning 3–5; model lessons 1–2; safety in science 7–8; sequencing lessons for 26–27; teacher leadership 3; three-legged model 3
Ahtee, M. 108
Andersson, B. 108
assessment 25, 52, 53, 88; active meaning-making 2, 112; centered classrooms 36; formative assessments 2, 25, 89; "Ice Cold Lemonade" probe 16–19, 24, 25; integrating formative and summative 25–26, 89; mixing water formative assessment probes 20, 22, 25; effective 112–115; "Sink or Float" probe 47, 49; summative assessment 24, 26, 89; "Watermelon and Grape" probe 44–45, 49, 51, 52, 53
Austin, M. 38

backward lesson planning 63, 93, 103–106; designing learning experiences 65–66; interdisciplinary benefits 64; practical steps 64; strategies supporting evidence-based claim construction 65; UbD framework 63; *see also* lesson planning
Bannister effect 118–120
Bannister, R. 118
belief-driven instructional change 121–124
Belinger 62
Ben-Zvi, R. 112
brain: development and education 34–35; reactive 35; thinking 34–35
Briggs, H. 108
Brown, P. 103

chemical changes 108, 110, 112
classroom science safety 7–8
cognitive science 35; assessment-centered approach 36; components and attributes of student learning 37; early learning cognition 32; knowledge-centered approach 36; learner-centered approach 35; metacognition 71–73, 74
collective efficacy 126
conceptual understanding 2, 53, 68, 111–112, 124, 128; elaborative explorations 70; enhancement

activities 69; enhancing understanding 69; incorporating readings, discussions, and lectures 69–70; integrating instructional strategies 69–70; NGSS frameworks and vocabulary 69; role of scientific vocabulary 69; transfer learning 70
critical inquiry 1
crosscutting concepts (CCCs) 5, 7, 27, 28, 55, 91
curriculum design 37–39

designing learning experiences 65–66
developmental psychology 32–33
didactic instruction 124
disciplinary core ideas (DCIs) 5, 7, 27, 28, 55, 91
discrepant events 67
Dweck, C. S. 33

early learning cognition 32
effective science curriculum-based learning 3–5
elaboration activities 27, 47, 70, 87, 110–112
elevator pitch 124, 128–129
El Guerrouj, H. 119
Elmore, R. 3
energy transfer *see* thermal and kinetic energy transfer
enhancement activities 26, 54, 69, 70, 109–112
evidence-based claims 25, 38, 53, 54, 63–64, 74, 85, 88–89; centering lesson design on 65; lessons on reactions with 105–106; standards checklist **94–102**; strategies to support 65
evidence-based science lessons 93, 103–106
Explore-Before-Explain 1–2, 13, 73; activating students' ideas 73; enhancement and introduction of concepts 74; hands-on exploration and evidence-based claims 74; helicopter fall time 78–79, 90; impact on professional practice 128; lesson planning 92; metacognition 74; mindset 123–124, 126, 128–129; practical template for instructional design 114–115; in practice 13; predict-and-test activity on buoyancy 44–55; reflection 74; research-informed instruction 30; rethinking practices 57–60; sequencing science instruction 13, 75; thermal energy model lesson 14, 15

fall time *see* helicopter fall time
"fixed mindset" 33
float test *see* predict-and-test activity on buoyancy
formative assessments 2, 25, 89
Frameworks 5–7
Freeman, S. 33
Fullan, M. 38, 63, 76, 93

"golden age" of science education reform 129
"growth mindset" 33, 117–120

hands-on activity 25, 70, 73, 93, 103, 122
Harris, D. 17
Hattie, J. 2, 71
helicopter fall time 78; activating prior knowledge 88; active meaning-making 78–79; contextualized learning 88; enhancement activity 86; exploration and evidence-based claims 88–89; force diagram exploration 86–87; guiding questions for reflection 79; inquiry planning wheel 80, *81*; integrating formative and summative assessments 89; investigating 80–87; just-in-time learning 89; materials needed 79; model-based reasoning and inquiry learning 86–87; model lesson 79, 88; safety notes 80; for secondary students 86–87; sequencing science

instruction 90; Standards in practice 90–91; student ideas 82; student-led inquiry into forces and motion 80–87; students' data **84, 86**; students evidence-based claims 85; for students in grades 3–5 and beyond 86; teacher enhancements 89

high-quality professional learning 4

Holding, B. 108

"Ice Cold Lemonade" probe 16, 24, 25; core activity 17–18; idea ownership and scientific modeling 18–19; observation to understanding 19; student ideas 16, 17; teacher perspectives 17, *18*; *see also* thermal and kinetic energy transfer

innovative thinking 58–60

instructional core 3

instructional design 42, 61, 114; essential questions for 107; innovation in 57–60; practices 121–122; principles 117; steps to revitalize 114, **115**

instructional programming 3

instructional strategies 69

intuitive inquiry skills 31, 39

inventive approach 58, 60

iterative approach 58, 60

just-in-time learning 54, 89

Keeley, P. 67, 103

learning: active 33, 68, 90, 123; high-quality professional 4; inquiry-based 65; just-in-time 54, 89; metacognitive aspect of 53; modern 31–32; passive 123; reflective 71–73; rethinking 30–39; targets 127–128; three-dimensional teaching and 6; transfer 70; visible 2

learning, rethinking 30; brain development and education 34–35; centering learning through sequenced instruction 38; cognitive science 35–36; curriculum design 37–39; developmental psychology 32–33; early learning and active exploration 32–33; enhancing achievement 39; 5E instructional model 37, 38; leveraging children's natural scientific thinking 31–32; modern learning 31–32; neurosciences 34–35; science education research 37–39

lectures 9, 69–70, 123, 124

lesson planning 61, 75–76, 92; activating students' ideas 66–68, 106; active meaning-making 114; activities to teach ideas 104–105; aligning enhancement explanations with standards 110; anchoring learning in students' lives 107; backward lesson planning 93, 103–106; chemical changes 108–109, 111–112; conceptual understanding 68–71, 111–112; elaboration activities 110–112; enhancement activities 109–112; evidence-based science lessons 93, 103–106; *Explore-Before-Explain* approach 73–75, 114–115; facilitating reflection and scientific claim revision 113; hands-on activities 93; instructional shifts 62–63; lessons on reactions 105–106; metacognition 71–72; "need-to-know" situation 106; planning backward 63–66; planning, essential elements, and activity example template **115**; promoting reflection on learning 71; reflective assessment 112–115; student evidence-based claims standards checklist **94–102**; students' ideas 107–109, 111–112; teacher-centered approach 92; traditional 75; tried-and-true activities 104

Mazur, E. 123

McTighe, J. 67

metacognition 71–73, 74
mindset 118, 126; *Explore-Before-Explain* mindset 123–124, 126; growth mindset in education 117–120; teacher mindsets 121–124
mixing water formative assessment probes 20, 22, 25
model lessons 1–2, 62, 127, 128; fall time 79; sinking and floating 43; thermal and kinetic energy transfer 14
modern learning 31–32
"More A–More B" 45

National Science Teaching Association (NSTA) 8
"need-to-know" situation 106
neurosciences 34–35
Next Generation Science Standards (NGSS) 3, 5; frameworks for vocabulary and concepts 69; three dimensions of **6**
notice and wonder questioning routines 66–67, 89, 107

Osborne, R. 82

passive learning 123
performance expectations (PEs) 64
Petrilli, M. 129
phenomena-based teaching 66
practices, rethinking 57–60
predict-and-test activity on buoyancy 44, 74; activating prior knowledge 52; common student ideas 45; constructivist approach 52; evidence-based student reasoning 45–49; guiding questions for reflection 43; incorporating assessments 53; investigating mass, volume, and buoyancy 49–51, **50**; just-in-time learning 54; materials needed 43; model lessons 43; "More A–More B" 45; observation to explanation 53; for primary students 45–49; relevancy of assessment probe 52; safety notes 44; for secondary students 49–51; sequencing science instruction 54–55; Standards in practice 55; students' evidence-based claims 53, 54; teacher enhancements 53–54; "Watermelon and Grape" probe 44–45, 49, 51, 52, 53
probing questions 67, 81, 113; assessment probe 52, 88; formative assessment probes 2, 25, 89; "Ice Cold Lemonade" probe 16–19, 24, 25; mixing water formative assessment probes 20, 22, 25; "Sink or Float" probe 47, 49; "Watermelon and Grape" probe 44–45, 49, 51, 52, 53
productive discourse 66–67
professional learning communities (PLCs) 1, 5

"reactive brain" 35
reflection 14, 43, 72, 74, 128; active meaning-making 2; reflective assessment 112–115; reflective learning 71–73
research-based instructional sequencing 42, 44–45
rethinking *see* learning, rethinking

safety in science 7–8
science and engineering practices (SEPs) 5, 7, 27, 28, 55, 91
science curriculum-based learning, effective 3–5
Science Curriculum Improvement Study (SCIS) curriculum 37
science education research 37–39
scientific vocabulary 20, 69
sinking and floating *see* predict-and-test activity on buoyancy
"Sink or Float" probe 47, 49
Solomonidou, C. 108
Standards-driven lessons 128
Stavridou, H. 108
summative assessment 24, 26, 89

synapses 34
synaptic pruning 34

targeted instruction 74
thermal and kinetic energy transfer 14, 15, 24; activating knowledge 24–25; demonstrations to real-world applications 20–24; for elementary students 19–20; enhancement activity 23–24; everyday experiences and energy concepts 25; formative assessment probes 25; guiding questions for reflection 14; "Ice Cold Lemonade" probe 16–19, 24, 25; integrating formative and summative assessments 25–26; investigating temperature mixing 22–23, 23; kinetic energy and water density activity 23–24; learning scientific vocabulary 20; materials needed 15–16; mixing water formative assessment probes 20, 22, 25; revising ideas through evidence 19–20; for secondary students 20–24; sequencing lessons 26–27; Standards in practice 27–28; student-centered inquiry 25; student misconceptions and cognitive development 21; students' evidence-based claims 25; summative assessment 26; teacher enhancements 26; vignette 16–19
"thinking brain" 34–35
three-dimensional teaching and learning 6
three-legged model 3
transfer learning 70
tried-and-true activities 104

Understanding by Design (UbD) framework 63, 74; *see also* backward lesson planning

Varjola, I. 108
visible learning 2
Vogelezang, M. 108

"Watermelon and Grape" probe 44–45, 49, 51, 52, 53
Willis, J. 67

For Product Safety Concerns and Information please contact our EU representative GPSR@taylorandfrancis.com
Taylor & Francis Verlag GmbH, Kaufingerstraße 24, 80331 München, Germany